CAMBRIDGE BIOLOGICAL STUDIES

General Editor

C. H. WADDINGTON

MECHANISMS OF
BIOLOGICAL OXIDATIONS

MECHANISMS OF
BIOLOGICAL OXIDATIONS

By

DAVID E. GREEN
SENIOR BEIT MEMORIAL RESEARCH FELLOW
INSTITUTE OF BIOCHEMISTRY
UNIVERSITY OF CAMBRIDGE

CAMBRIDGE
AT THE UNIVERSITY PRESS
1940

CAMBRIDGE
UNIVERSITY PRESS

University Printing House, Cambridge CB2 8BS, United Kingdom

Cambridge University Press is part of the University of Cambridge.

It furthers the University's mission by disseminating knowledge in the pursuit of education, learning and research at the highest international levels of excellence.

www.cambridge.org
Information on this title: www.cambridge.org/9781107560109

© Cambridge University Press 1940

First published 1940
First paperback edition 2015

A catalogue record for this publication is available from the British Library

ISBN 978-1-107-56010-9 Paperback

To

M. DIXON

F. G. HOPKINS

and

D. KEILIN

CONTENTS

CHAPTER I

Introduction: General properties of oxidation enzymes

There are three fundamental problems in connection with cellular oxidations, viz. which substances are oxidized, how are they oxidized and how is the energy so derived utilized by the cell? Unfortunately our knowledge of how energy liberated in oxidation processes is transformed into other forms of energy, such as osmotic, electrical, etc., is very meagre. Biochemists have so far concerned themselves principally with the problems of intermediary metabolism and mechanism of reaction, and for lack of experimental methods have been obliged to shelve the problem of energy transmission.

The mechanisms of cellular oxidation are the main concern of this book. The greater part of our knowledge of mechanism is derived from experiments on a few animal tissues, special yeasts and bacteria. The question may well be raised, to what extent is the biochemist justified in reasoning from one cell to another? May we assume that mechanisms which hold for specific yeasts and bacteria apply as well to moulds, fungi, higher plants, and bacteria in general? There is no unequivocal answer as yet to this question. Cells of every description have some chemical equipment in common. For example, catalase, cytochrome, riboflavin and the pyridine nucleotides are found in bacteria, plants and animal tissues. It is tempting to interpret these elements of universality as proof that the mechanisms of oxidation are more or less similar for all types of cell. There is, however, little to be gained by such an over-simplification. Practically nothing is known about the oxidative mechanisms of the vast majority of cells. It remains for future experimentation to decide whether oxidations in different types of cells are merely variations on the same theme.

Enzymic catalysis is the motif of all cellular oxidations. The oxidation of foodstuffs or metabolites is not a process which occurs spontaneously; it requires the presence of highly specific protein molecules known as enzymes. For example, alanine in aqueous solution at 38° is extremely resistant to oxidation. However, in presence of a specific enzyme extracted from liver or kidney, alanine is rapidly oxidized by molecular oxygen. The enzyme is said to catalyse the oxidation of alanine. The properties of these enzymes will be considered in detail in later sections and for the moment the question of how enzymes work will be deferred. It is by virtue of the intracellular catalysts that the cell is able to carry out oxidation of foodstuffs with extraordinary velocities at low temperatures, and without the use of vigorous oxidizing agents.

There is no universal oxidizing agent for all cells. Molecular oxygen is the ultimate agent in all aerobic organisms. However, anaerobic organisms perform a multitude of oxidations in the complete absence of oxygen, and even aerobic organisms are not entirely dependent upon oxygen as the oxidizing agent. The study of the mechanisms of biological oxidations has elucidated the existence of oxidizing agents other than oxygen. The details of any cellular oxidation-reduction process will be found to depend upon the particular oxidizing agent which comes into play.

We may assume that the oxidation of any substance must involve one of the following four types of chemical change (or combinations thereof):

(1) Loss of electron,

$$\text{e.g. } Fe^{++} \underset{\text{reduction}}{\overset{\text{oxidation}}{\rightleftharpoons}} Fe^{+++} + \text{ⓔ}$$

(2) Loss of hydrogen,

(3) Gain of oxygen,

$$\text{e.g. } CO + \tfrac{1}{2}O_2 \rightarrow CO_2$$

(4) Addition of water with loss of hydrogen,

$$\text{e.g. } CCl_3CHO + H_2O \rightarrow CCl_3CHO\,.\,H_2O$$

$$CCl_3CHO\,.\,H_2O \rightarrow CCl_3COOH + \text{(2H)}$$

The oxidizing agent is referred to as an electron acceptor in case (1) and as a hydrogen acceptor in case (2). Needless to say, neither a free electron nor atomic hydrogen is liberated in solution; the transfer is made directly to the appropriate acceptors. The circles around the symbols for the electron and atomic hydrogen indicate this transfer. In oxidations involving gain of oxygen either molecular oxygen or some oxide can act as the oxidizing agent. Case (4) is essentially a combination of cases (2) and (3).

In the present state of our knowledge it is very often a matter of convention whether a particular reaction shall be considered as one involving a loss of electrons or a loss of hydrogen. The net result is the same, although the postulated mechanisms are different. The oxidation of many organic compounds entails a simultaneous transfer of a hydrogen atom and an electron. In cellular oxidations we are dealing with all these four types of oxido-reductions, though mechanisms involving a transfer of hydrogen are most common.

The reduction of substances involves the converse processes to those of oxidation, i.e. gain of electrons, gain of hydrogen or loss of oxygen. Oxidation and reduction are simultaneous, mutually dependent processes. The problem of how foodstuffs are oxidized is therefore inextricable from the problem of how oxidizing agents are reduced.

Before proceeding to the systematic study of biological oxidations a survey of the properties and characteristics of the amino-acid oxidase will be made. It is not that this enzyme is unusually important, but rather that no proper appreciation of biological oxidations is possible without some familiarity with the properties of oxidation enzymes. The general properties of the amino-acid oxidase may be assumed to parallel those of other oxidation enzymes.

The enzyme is easily prepared from the kidneys of ox, pig or sheep by the following method (1). The kidney is finely minced and mixed with 7 vol. of acetone. The mixture is stirred and filtered. The residue is again mixed with acetone and filtered off. The almost dry residue is collected, spread in a thin layer and allowed to dry at room temperature. The dry powder is rubbed up with *ca.* 20 vol. of buffer *p*H 9 and the mixture is centrifuged. The somewhat opalescent supernatant fluid contains the active enzyme.

A mixture of $d(-)$-alanine and the enzyme absorbs molecular oxygen. Chemical analysis shows that coincident with the oxygen uptake alanine disappears whilst pyruvic acid and NH_3 appear. The following are the stoichiometric relations:

$$CH_3CHNH_2COOH + O_2 + H_2O \rightarrow CH_3COCOOH + NH_3 + H_2O_2$$

There is a transfer of two atoms of hydrogen from alanine to oxygen,* the former being oxidized to pyruvic acid and NH_3, the latter being reduced to H_2O_2. If the enzyme is destroyed by boiling the solution for a few seconds no reaction between alanine and oxygen takes place. It is clear that the enzyme catalyses a reaction which is not spontaneous.

The catalytic power of the amino-acid oxidase is not confined to $d(-)$-alanine. Many other α-amino acids of the same optical series, i.e. of the d-configuration, are also oxidized by oxygen in presence of the enzyme (1). However, the oxidation of simple amines, α-amino acids of the l series, diamines, fatty acids and any other group of organic compounds, is not catalysed by the enzyme. Thus the amino-acid oxidase is specific not for a particular amino acid but for the group of α-amino acids stereochemically related to $d(-)$-lactic acid, i.e. the non-naturally occurring isomer. Not all oxidation enzymes show *group* specificity. The malic enzyme of animal tissues, for example, catalyses only the oxidation of $l(-)$-malic acid. This enzyme by contrast shows specificity for an individual *compound*. In general, oxidation enzymes exhibit a high degree of specificity for their substrates, i.e. the substances whose oxidation is catalysed.

The speed of oxidation of alanine by molecular oxygen in presence of a given amount of enzyme and a fixed partial pressure

* The oxidation of alanine proceeds in two stages:

$$CH_3CHNH_2COOH \rightarrow CH_3C:NHCOOH + \textcircled{2H}$$
$$CH_3C:NHCOOH + H_2O \rightarrow CH_3COCOOH + NH_3$$

The first and catalytic stage is the dehydrogenation of alanine to the corresponding imino acid. The second and spontaneous stage is the hydrolysis of the imino acid to the corresponding ketonic acid with liberation of NH_3.

of oxygen gas depends for a given temperature upon two factors:
(1) the hydrogen-ion concentration and (2) the concentration of
alanine. Fig. 1 shows how the rate of oxidation is affected by the
*p*H. Below *p*H 7 and above *p*H 10 the rate falls off sharply to
zero. The maximum activity is shown at *ca.* *p*H 9. Little importance can be attached to the maximum *p*H, since the value
depends on the particular buffer system employed. The *p*H curve
with one buffer system rarely parallels that of another. But regardless of which buffer system is employed no activity is found
at *p*H values lower than 6 or higher than 11. It is not entirely

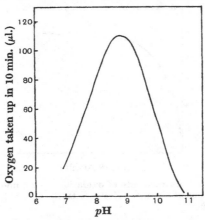

Fig. 1. *p*H curve of the *d*-amino-acid oxidase. Substrate *dl*-alanine. (Krebs
[1935]. *Biochem. J.* **29**, 1620.)

clear why activity disappears in these ranges of hydrogen-ion
concentration. No doubt destruction of the enzyme plays some
role. But other factors such as the ionization of the enzyme and
the substrate are probably concerned. Fig. 2 shows how the rate
of oxidation depends on the concentration of alanine. The rect-
angular hyperbola relation is quite typical for enzymes in general.
The substrate concentration giving half the maximum velocity
(i.e. the Michaelis constant Km) obtained from the curve is
5×10^{-3} M. The value of the Km varies from enzyme to enzyme.
The equilibrium of compound formation between enzyme and
substrate determines the value of the Michaelis constant.

The velocity of oxidation of alanine is directly proportional
to the concentration of enzyme. Certain enzymes exhibit no
measurable activity below a critical dilution of the enzyme. This
phenomenon is usually attributed to a rapid rate of destruction
of the enzyme in the course of its catalytic function. If the rate
of destruction is high and the amount of enzyme is small, complete destruction of the enzyme sets in before any measurable
oxidation takes place.

The rate of oxidation of alanine by oxygen in presence of the
enzyme is linear only for about 20 min. and then gradually falls

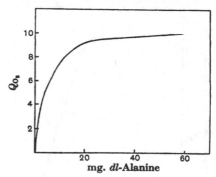

Fig. 2. Relationship between rate of oxidation by amino-acid oxidase and
concentration of *dl*-alanine. (Keilin & Hartree [1936]. *Proc. Roy. Soc.*
B, 119, 114.)

off. This is quite a general phenomenon among oxidation
enzymes. It is possible to rule out such factors as inhibition of
the reaction by the reaction products or mass-action equilibria.
Experiment shows that in the main the falling off in rate is due
to the destruction of the enzyme. There is little information
available as to why oxidation enzymes in particular exhibit such
fragility in the course of their catalytic function in reconstructed
systems, and it would be of interest to know the lifetime of an
enzyme in the intact cell.

The amino-acid oxidase can be reversibly resolved into a
colourless protein and a yellow substance of comparatively low
molecular weight, which we shall refer to as the prosthetic
group [2]. Neither portion alone can catalyse the oxidation of

alanine. The catalysis requires the simultaneous presence of the two components. Fig. 3 shows how the rate of oxidation of alanine depends upon the concentration of the yellow prosthetic group (flavinadenine dinucleotide) for a given amount of the colourless protein. It is obvious from the curve that the protein forms a compound with the prosthetic group in a definite stoichiometric proportion and that the protein-prosthetic group compound dissociates under the conditions of the experiment. Dissociation is indicated by the fact that the asymptotic maximum activity value is not reached abruptly. In neutral or

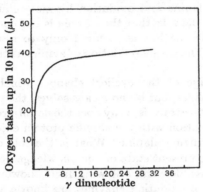

Fig. 3. Relationship between rate of oxidation of alanine by amino-acid oxidase and concentration of flavinadenine dinucleotide. (Warburg & Christian [1938]. *Biochem. Z.* **298**, 150.)

slightly alkaline solutions the equilibrium is in favour of dissociation.

Both components of the amino-acid oxidase are highly specific. The protein cannot be replaced by any other protein. Similarly, the flavinadenine dinucleotide prosthetic group cannot be replaced by compounds very similar in chemical structure, such as riboflavinphosphate or adenylic acid(2). It is therefore obvious that both components of the amino-acid oxidase are essential for catalytic activity.

A concentrated solution of the highly purified oxidase is yellow in colour. On addition of alanine to the oxidase solution the colour is bleached(3). On shaking vigorously with air the colour is restored. The colour of the oxidase is due entirely to the flavin moiety. It is well known that reducing agents bleach the colour of flavin solutions and that reduced flavins are rapidly oxidized by molecular oxygen with production of H_2O_2. The oxidation of alanine by molecular oxygen may now be described by the following equations:

alanine + oxidase $\rightarrow$ pyruvic acid + NH_3 + reduced oxidase,

reduced oxidase + O_2 $\rightarrow$ oxidase + H_2O_2.

The oxidase catalyses the reaction by undergoing a cycle of reduction by alanine and oxidation by molecular oxygen. An interesting corollary is that the oxygen is not concerned in the oxidation of alanine but is required only to reoxidize the reduced oxidase. The actual oxidation of alanine is the prerogative of the oxidase.

The knowledge of the cyclical changes which the oxidase undergoes simplifies, but by no means solves, the problem of the mechanism of the catalysis. Why does the flavin-prosthetic group when in combination with the specific protein enjoy the unique property of oxidizing alanine? What is the role of the specific protein? In the present state of our knowledge no complete explanation is possible. The following facts, however, point the way. Families of oxidation enzymes are known which have in common the same prosthetic group. Each member of such families differs from all the others in its specific protein and in the nature of its catalysis. In other words the specific protein portion determines the nature of the catalysis. From this fact it follows that the protein must be structurally constituted so as to exert some special influence on the substrate. What is the nature of this influence? Consider the case of the succinic enzyme which catalyses the oxidation of succinic acid ($COOHCH_2CH_2COOH$) to fumaric acid. Malonic acid ($COOHCH_2COOH$), which is the lower homologue of the dicarboxylic acid series, is not attacked by the enzyme. The enzymic oxidation of succinic acid is strongly inhibited by low concentrations of malonic acid (4) and can be completely abolished by high concentrations. Such striking inhibitory effects are not produced by any other carboxylic acid yet tested. The following explanation is given for this phenomenon of competitive inhibition. The actual process of oxidation is preceded by the formation of a substrate-enzyme compound. A special group in the protein molecule (usually referred to as the active group) is concerned in this linkage. To return to the example of the succinic enzyme, the "active" group is capable of combining not only with succinic acid but also with the homologous compound malonic acid. But whereas succinic acid undergoes oxidation malonic acid remains unchanged. The inhibition by malonic acid is therefore explicable in terms of a competition between malonic acid and succinic acid for the active group of the enzyme. This phenomenon of competitive inhibition is general for all oxidation enzymes as yet tested and is the best available proof for the assumption that the substrate combines with the enzyme, and that a special group of the enzyme is concerned in this combination.

The next problem is to explain why the substrate undergoes oxidation when combined with the enzyme. On that point experiment is silent though theory is eloquent. It is stated that the enzyme "activates" the substrate or that a "strain" is imposed on the substrate molecule. These are terms of ignorance and merely express the experimental fact that the substrate undergoes an oxidation in presence of the enzyme which it does not undergo spontaneously. The nature of activation or straining of the substrate molecule is the spearhead of present research in biological oxidations.

The description of the flavinadenine dinucleotide portion of the amino-acid oxidase as the prosthetic group does not imply that it is the only catalytic portion of the molecule, and that the protein is merely ballast. In some cases, as in that of the amino-acid oxidase, the prosthetic group undergoes a cycle of oxidation and reduction. In other cases, as in that of peroxidase, the prosthetic group undergoes another type of cyclical change. By prosthetic group we shall mean the nucleus of comparatively low molecular weight with which the specific protein is combined or is in association. This nucleus may be flavin dinucleotide, Fe porphyrin, copper, pyridine dinucleotide or thiamine.

The classification of oxidation enzymes may be based (1) on the chemical nature of the substrates, (2) on the components of the complete oxidation systems, and (3) on the chemical nature of the prosthetic group. From the point of view of ease of presentation the third basis of classification seems most satisfactory and will be used in this book whenever possible. In Chapters II–VI we shall consider oxidation enzymes with known prosthetic groups. Chapters VII–IX deal with enzymes whose prosthetic groups if any are still unknown.

REFERENCES

(1) Krebs [1935]. *Biochem. J.* 29, 1620.
(2) Warburg & Christian [1938]. *Biochem. Z.* 298, 150.
    Straub [1938]. *Nature, Lond.*, 141, 603.
(3) Negelein & Brömel [1939]. *Biochem. Z.* 300, 225.
(4) Quastel & Wooldridge [1928]. *Biochem. J.* 22, 689.
    Quastel & Wheatley [1931]. *Biochem. J.* 25, 117.

Chapter II

Fe porphyrin protein enzymes

Haemoglobin is the most familiar example of a class of pigmented compounds widely distributed in nature. These compounds with diverse functions, such as the transport of oxygen and the catalysis of oxidative and photosynthetic processes, have in common a structural unit known as porphyrin. The work of Fischer and others has firmly established that the framework of the porphyrins consists of four pyrrole rings linked together into a more complex ring system by four methin (CH) bridges:

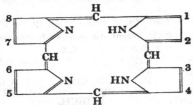

Skeleton formula of porphyrin

Porphyrins easily combine with various metals, such as Fe, Cu, Co, Mg, Mn, Zn, etc. (1), giving rise to a series of metallo-porphyrins in which the metals are bound to the N atoms of the pyrrole nuclei. In nature only Fe, Cu and Mg have been found in combination with porphyrins:

Fe: haemoglobin, chlorocruorin, helicorubin, catalase, peroxidase, cytochrome.

Cu: turacin.

Mg: chlorophyll.

It is significant that oxidation catalysts which are known to be porphyrin compounds all contain Fe. The metal of Fe porphyrin compounds can exist either in the ferrous or ferric condition.

At the carbon atoms numbered 1 to 8 in the pyrrole rings of porphyrin, side chains of various types are encountered. Table I contains a summary of the side chains of the more important porphyrins. Each individual porphyrin has many isomeric forms (denoted by Roman numerals) depending on the arrangement of the substituent side chains. For the purposes of the present discussion we shall consider only the isomers of the different porphyrins which are configuratively identical with the natural porphyrins. Fischer & Klarer (46) have synthesized aetioporphyrin III and established its constitution. By suitable degradation methods it is possible to convert protoporphyrin, the natural porphyrin of haemoglobin, helicorubin, catalase and peroxidase, into aetioporphyrin. Degradation of chlorophyll, by hydrolysing the ester groups and removing Mg, also yields aetioporphyrin III identical with the synthetic product (Fischer & Willstätter (46,47)). These observations serve to link together by direct experimental methods the porphyrins of the haemoglobins and chlorophylls.

Table I. *The substituent side chains of the porphyrins*

(Pryde [1931]. *Recent Advances in Biochemistry.*)

	1, 3, 5, 8	2, 4	6, 7
Aetioporphyrin	CH_3	C_2H_5	C_2H_5
Deuteroporphyrin	CH_3	H	CH_2CH_2COOH
Protoporphyrin	CH_3	$CH:CH_2$	CH_2CH_2COOH
Mesoporphyrin	CH_3	C_2H_5	CH_2CH_2COOH
Haematoporphyrin	CH_3	$CHOHCH_3$	CH_2CH_2COOH
Coproporphyrin	CH_3	CH_2CH_2COOH	CH_2CH_2COOH
Uroporphyrin	CH_3	$CH_2CH(COOH)_2$	$CH_2CH(COOH)_2$

Haemoglobin, although not an oxidation catalyst, serves as a helpful guide to the properties of oxidation enzymes of the Fe porphyrin class. The red pigment can be reversibly resolved (2,3) into a colourless protein, globin, and a coloured prosthetic group, ferroprotoporphyrin. It is important to note that the Fe atom is attached to the pyrrole nuclei by both electrovalencies and covalencies. The mode of attachment of Fe in different porphyrin compounds has several variations.

haem

Haem (ferroprotoporphyrin) is oxidized by molecular oxygen to haematin (ferriprotoporphyrin)[3]. Conversely, haematin can be reduced with hyposulphite or cysteine to haem, the ferro form. Both haem and haematin can combine with certain proteins and bases to form haemochromogens and parahaematins respectively[4]. None of these compounds has the property of binding molecular oxygen in the same way as haemoglobin. Hill & Holden[2] and Anson & Mirsky[3] have shown that haem and haematin combine with globin to form haemoglobin and methaemoglobin respectively. Haemoglobin stands in relation to methaemoglobin as ferrous to ferric. The transition from haemoglobin to oxyhaemoglobin does not involve a change in the valency of iron and this process can be described as oxygenation in contradistinction to oxidation. According to Pauling et al.[5] the bonds attaching the iron atom to the pyrrole nuclei in haemoglobin are essentially ionic, whereas in the haemochromogens and in oxyhaemoglobin the bonds are essentially covalent. It has long been recognized that the compound formed by the union of globin and haem is quite distinct in many properties from the haemochromogens.

Haemoglobin can combine with CO in the ratio of one molecule of CO per atom of Fe. Methaemoglobin can combine in the ratio of one molecule of reagent per atom of Fe with fluoride, cyanide, hydrosulphide[6], azide[7], hydrogen peroxide and ethyl hydrogen peroxide, but not with CO. For purposes of simplicity we may consider methaemoglobin as a univalent base (HbOH) which dissociates in neutral or acid solution[8]:

$$HbOH \rightleftharpoons Hb^+ + OH^-$$

Replacement of the hydroxyl group with various anions gives rise to salt-like compounds such as HbF, HbCN, HbSH, HbN$_3$, etc. There is always an equilibrium between the salt and the base, e.g. in the case of the fluoride:

$$HbF + OH^- \rightleftharpoons HbOH + F^-$$
$$HbF \rightleftharpoons Hb^+ + F^-$$

The spectroscopic and chemical properties of these methaemoglobin derivatives are quite characteristic.

In the haemochromogens one molecule of the Fe porphyrin is combined with two molecules of base (9). Haemochromogens in neutral solution readily oxidize in air to the corresponding parahaematins. Reduction with hyposulphite brings about the reverse transformation. Haemochromogens can combine with CO, one molecule of nitrogenous base being replaced by one of CO:

$$\text{porphyrin Fe}^{++}\!\!\!\begin{array}{c}\text{base}\\ \diagup \\ \diagdown \\ \text{base}\end{array} + \text{CO} \rightleftharpoons \text{porphyrin Fe}^{++}\!\!\!\begin{array}{c}\text{CO}\\ \diagup \\ \diagdown \\ \text{base}\end{array} + \text{base}$$

Irradiation with strong light dissociates the CO complex and thereby shifts the equilibrium to the left.

Among oxidation enzymes of the Fe porphyrin class catalase and peroxidase resemble haemoglobin, whereas the cytochromes are more akin to the haemochromogens in their spectroscopic and chemical properties. All the members of the class have different proteins. Catalase and peroxidase have the same prosthetic group as that of haemoglobin, viz. protoporphyrin. Cytochrome *c* contains a porphyrin which is similar to but not identical with protoporphyrin. The nature of the porphyrins in cytochromes *a* and *b* is still unknown.

Catalase

Catalase is an enzyme which catalyses the decomposition of hydrogen peroxide into water and molecular oxygen:

$$2H_2O_2 \rightarrow 2H_2O + O_2$$

The decomposition of hydrogen peroxide by living cells was ascribed to a specific enzyme by Loew in 1901. The enzyme is widely distributed in animal and plant tissues and in microorganisms. Zeile & Hellström in 1930 (10) were the first to observe that the absorption spectrum of purified preparations of catalase is similar to that of Fe porphyrin compounds and that the activity of the enzyme closely parallels the concentration of Fe porphyrin present in the preparation. Stern (11) denatured catalase with acidified acetone and isolated protoporphyrin IX, which is the unmodified porphyrin of haemoglobin. Sumner & Dounce (12) observed in the acetone extract a blue substance in addition to the porphyrin. Later Lemberg *et al.* (43) identified this blue substance as the iron-free bile pigment biliverdin. According to Lemberg *et al.* (43) biliverdin does not occur in catalase as such but is liberated by acid from an iron-containing precursor which is intermediate in constitution between a closed ring Fe porphyrin and an open chain, iron-free bile pigment (cf. Lemberg (13, 43) for a discussion of the chemistry of verdohaemochromogen and related compounds).

haemochromogen   verdohaemochromogen

biliverdin

The ratio of biliverdin to Fe porphyrin found after denaturation of the enzyme was uniformly 1 : 3 (43). Since the molecular weight of catalase is 248,000, and since the total iron is *ca.* 0·1 % (12,39), it follows that there must be three molecules of Fe porphyrin and one molecule of bile pigment Fe porphyrin per molecule of catalase.

Sumner & Dounce (12) have succeeded in crystallizing catalase obtained from ox liver. Their method of preparation deserves to be a classic from the point of view of its simplicity and effectiveness. 300 g. well-minced liver is mixed with 400 ml. 35 % dioxane. The insoluble material is filtered off. To every 100 ml. of filtrate 20 ml. of dioxane are added. After the solution has stood for 12–24 hr. at 0° the precipitate is filtered off. Dioxane is again added to the filtrate (10·2 ml. to every 100 ml.) and catalase is now precipitated. The precipitate is resuspended in a small volume of water, diastase is added to hydrolyse residual glycogen

16   FE PORPHYRIN PROTEIN ENZYMES

and the suspension is dialysed against distilled water. After a short time crystals of catalase (prisms or plates) appear in the dialysis sac. The enzyme can crystallize either in the form of prisms and plates, or as needles. The two crystalline forms differ somewhat in solubility. It is possible, however, to convert one form into the other by altering the conditions under which crystallization takes place. The crystals look black, yellow or silvery according to the illumination. Repeated crystallizations increase the catalatic activity up to an asymptotic value. Sumner & Dounce report that their best preparations are homogeneous in both ultracentrifugal and cataphoretic fields.

One molecule of catalase decomposes $4 \cdot 4 \times 10^4$ molecules of H_2O_2 per sec. at $0°$ (14). One mg. of catalase iron produces 2740 l. of oxygen from H_2O_2 per hr. at $0°$. For determining these values it is not necessary to have a homogeneous preparation of catalase, since the amount of enzyme can be estimated spectrophotometrically with great accuracy even in partially purified preparations. The spectrophotometric method estimates only iron combined with porphyrin and is therefore more reliable than the chemical method, which estimates total iron.

A strong catalase solution examined with a Hartridge reversion spectroscope shows three bands (14): I, 629 mμ.; II, 544 mμ.; III, 506·5 mμ.* These bands can be modified either in position or intensity by means of reagents which are known to poison catalase activity. Table II contains a summary of the bands of

Table II. *Visible absorption bands of catalase and derivatives*
(Keilin & Hartree [1936]. *Proc. Roy. Soc.* B, **121**, 173.)

	I	II	III
Catalase	629·5 mμ.	544 mμ.	506·5 mμ.
KCN catalase	595·5 mμ.	556·5 mμ.	
H$_2$S catalase	640·5 mμ.	587 mμ.	548 mμ.
NaF catalase	622 mμ.	597·5 mμ.	
NO catalase	577 mμ.	538·5 mμ.	
NaN$_3$ catalase	624 mμ.	544 mμ.	506·5 mμ.

catalase in presence of some of these reagents. It is obvious that there is an exact correlation between the effect of reagents on

* Sumner & Dounce (12) report the following positions of the bands: 627, 536, 502 mμ.

enzyme activity and their effect on the absorption spectrum of catalase. Addition of H_2O_2 to a strong catalase solution produces no change in the bands, although a violent evolution of O_2 ensues. If there is a combination between the substrate and the enzyme, there should be some appreciable change in the spectrum, providing the life period of the enzyme-substrate complex is not too short. The available evidence indicates that the excessively short life period of the complex explains why addition of H_2O_2 does not affect the absorption bands of catalase.

The analysis of the mode of action of catalase by Keilin & Hartree[17] has been accomplished principally by the use of inhibitors such as azide and hydroxylamine, which form definite compounds with catalase. These substituted catalases undergo reaction with H_2O_2 much more slowly than catalase itself. The possibility is thereby presented of analysing the various steps in the catalytic decomposition of H_2O_2.

If a small amount of H_2O_2 is added to a strong solution of azide-catalase the former is decomposed and during the process the solution turns distinctly red—the absorption spectrum of azide-catalase being replaced by a spectrum with two bands at 588 and 547 mμ. respectively[14]. When most of the H_2O_2 is decomposed the colour of the solution turns greenish-brown and the absorption spectrum of azide-catalase reappears. This experiment can be repeated several times with the same sample of azide-catalase. The shift from brownish-green to red on addition of peroxide can also be demonstrated with hydroxylamine-catalase.

The red compound can be shown to be a ferrous porphyrin compound by experiments of the following type. H_2O_2 is mixed with azide-catalase in an exhausted Thunberg tube[14]. The colour of the solution turns red and remains unchanged for a long time. On allowing air to enter the tube and shaking the solution, the colour returns to greenish-brown. On evacuation and washing the tube with nitrogen gas the red colour and the two-banded spectrum reappear. It is interesting that ferricyanide, which oxidizes all other ferrous porphyrin compounds, cannot replace molecular oxygen as the oxidant for ferro or reduced azide-catalase.

Another striking proof that the red compound is in the divalent or reduced form lies in its reaction with CO. The original greenish-brown azide-catalase does not combine with CO. But when reduced by H_2O_2 to the red form it rapidly combines with CO—the absorption bands of the CO compound being 10 mμ. nearer the blue end of the spectrum. In presence of oxygen the original azide-catalase is regenerated. Thus the reaction with CO is reversible. The evidence that CO combines only with the divalent or reduced forms of Fe porphyrin compounds is fairly decisive.

Keilin & Hartree[14] have compared the various inhibitors of the catalase reaction according to the effects they produce on the Fe porphyrin of catalase. The inhibitors may be separated into two categories: (1) inhibitors like KCN and H_2S, which prevent the reduction of ferric catalase by H_2O_2; (2) inhibitors like NaN_3, NH_2OH or NH_2NH_2, which do not prevent the formation of the ferro porphyrin but on the contrary stabilize it by preventing its reoxidation by molecular oxygen and in this way inhibit the catalase reaction.

The results of these various experiments led to the following formulation of the mode of action of catalase[15]:

$$4 \ Fe^{+++}_{(catalase)} + 2H_2O_2 \rightarrow 4 \ Fe^{++}_{(catalase)} + 4H^+ + 2O_2$$
$$4 \ Fe^{++}_{(catalase)} + 4H^+ + O_2 \rightarrow 4 \ Fe^{+++}_{(catalase)} + 2H_2O$$

H_2O_2 combines and reacts with catalase, the net result being the reduction of ferric catalase to the ferrous form and the oxidation of H_2O_2 to oxygen and water. That iron can be reduced by H_2O_2 is easily demonstrated by mixing a solution of ferric salt and $\alpha\alpha'$-dipyridyl with H_2O_2. The divalent iron formed in this reaction rapidly gives the red ferrous $\alpha\alpha'$-dipyridyl compound. If this formulation is correct we should expect that, by removing all traces of oxygen from a catalase solution, it should be possible either to reduce or completely to inhibit catalase activity. This has been shown experimentally by Keilin & Hartree[15] to be the case (cf. Fig. 4). The objection might be raised that in the reaction between H_2O_2 and catalase, molecular oxygen is formed. There should therefore be available one atom of oxygen for each atom of Fe, which is twice the equivalent amount necessary for oxidation. It appears that the reoxidation of ferro catalase can take place only when the oxygen tension exceeds a definite but by no means negligible oxygen tension, and that the amount of oxygen produced at the start of the catalytic decomposition is not sufficient to satisfy the minimum requirements of oxygen tension.

The Fe porphyrin of catalase is unique among Fe porphyrins in three respects: (1) its ferro form is not oxidized by ferricyanide; (2) its ferric form is not reduced by hyposulphite, and (3) it can

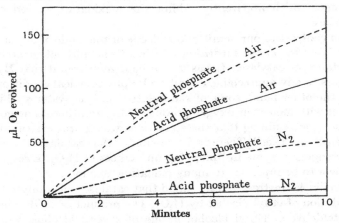

Fig. 4. The dependence of catalase activity on the presence of oxygen. (Keilin & Hartree [1938]. *Proc. Roy. Soc.* B, **124**, 397.)

be reduced only by H_2O_2. But otherwise the Fe porphyrin of catalase behaves as if it were combined with globin, e.g. the ferro form reacts with CO and the ferric form reacts with KCN, H_2S, NO, NaN_3, NaF, NH_2OH and NH_2NH_2.

The intermediary catalase-H_2O_2 complex has not as yet been demonstrated, but there seems to be some evidence that it does exist. The compound formed by catalase and ethyl-hydrogen peroxide (Stern(16)) is not an enzyme-substrate complex according to Keilin but a ferric-catalase compound similar to azide-catalase. Stern(16) finds that ethyl-hydrogen peroxide is decomposed in presence of catalase, yielding acetaldehyde and probably alcohol. But that occurs only when concentrated catalase solutions are employed and there is the possibility that extraneous Fe porphyrin may be responsible for this reaction. For example, methaemoglobin forms a definite compound with ethyl-hydrogen peroxide, which rapidly decomposes yielding acetaldehyde.

Some catalase preparations have been obtained which show a more or less marked light[15] sensitive inhibition by CO. Purified catalase preparations become sensitive to CO in presence of a very small concentration of azide, cysteine or glutathione, substances which apparently inhibit the reoxidation of reduced catalase.

In contrast to our detailed knowledge of the mode of action of catalase, we know surprisingly little of the physiological importance of catalase. It has yet to be demonstrated that H_2O_2 occurs to any appreciable extent under physiological conditions. On the other hand, some of the isolated enzyme systems which react with molecular oxygen produce H_2O_2, and there is no good reason for assuming that these reactions would proceed any differently in the cell. There is the possibility that catalase acts as a scavenging enzyme in "mopping up" traces of H_2O_2, a reagent known to be injurious to many oxidation enzymes.

Keilin & Hartree[17] have found that catalase can catalyse the oxidation of ethyl alcohol by H_2O_2. The product of oxidation is acetaldehyde. Ethyl alcohol is one of several alcohols which show this effect. H_2O_2 in high concentration inhibits the oxidation. The reaction takes place most effectively when there is a slow, continuous generation of H_2O_2 from peroxide derivatives such as BaO_2. The catalytic mechanism of the reaction is not yet clear.

There is a vast literature dealing with the kinetics and properties of catalase from different sources. The subject is fully treated in the monograph of Euler[18] in the section on catalase (written by Euler & Zeile jointly). Euler and his colleagues carried out the pioneer investigations on the properties and purification of catalase.

Peroxidase

Peroxidase is an enzyme widely distributed in plant tissues, which catalyses the oxidation of a variety of compounds by means of H_2O_2. Among animal tissues it is found in milk and possibly in spleen. Iodide, diamines such as o-phenylene diamine and benzidine, phenolic substances such as guiacum and p-cresol, aromatic mono-amines such as aniline, indicators and dyes such as phenolphthalein and leuco-malachite green are among those known to be attacked. H_2O_2 cannot be replaced as an oxidant by molecular oxygen, by organic peroxides (with the exception of monosubstituted products), or by oxidation-reduction dyes. A few substances of physiological interest [19] are known to be attacked by peroxidase, viz. tryptophan, ascorbic acid, adrenaline, tyrosine and histidine.

The chemistry of the many oxidations catalysed by peroxidase may be simple or complex, depending upon the substrate. The oxidations of nitrite and iodide to nitrate and iodine respectively are examples of simple, straightforward reactions. However, the oxidation of pyrogallol to purpurogallin is complex, for at least six consecutive reactions must be assumed to account for the net change (Willstätter [20]).

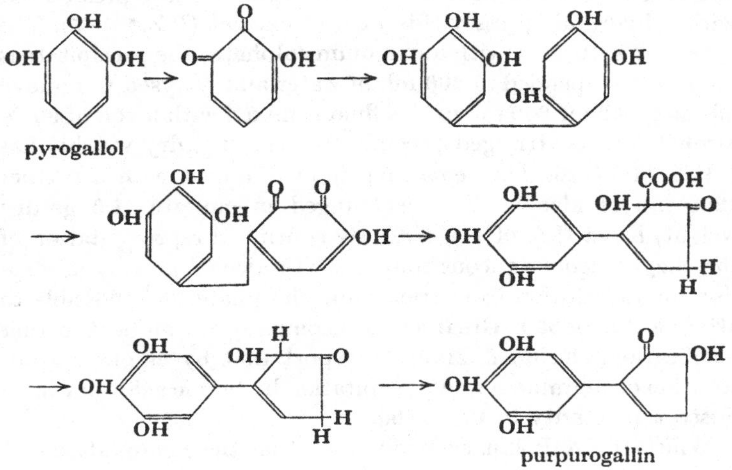

pyrogallol

purpurogallin

It is difficult to determine which reactions beyond the first are catalysed or spontaneous. The o-quinones which are formed from the benzenoid substrates are highly reactive and their subsequent reactions may not be catalysed. It is perhaps this reactivity of the oxidation products which explains why no simple statement of what peroxidase does can be formulated.

Practically all Fe porphyrins can act as pseudo-peroxidases, i.e. they catalyse the same type of reactions as peroxidase but never approach its catalytic efficiency (19). Peroxidase is thermolabile, whereas the pseudo-peroxidases found wherever Fe porphyrin occurs in tissues are all thermostable (22).

Willstätter and his colleagues are mainly responsible for the success which has attended the isolation of peroxidase. They developed the whole technique of purifying enzymes by adsorption and elution from specially prepared gels. This technique has proved to be one of the most powerful tools in enzyme isolation work.

Peroxidase is best prepared from the roots of non-cultivated horseradish by the combined methods of Elliott (19) and Keilin & Mann (23), which are extensions of the original methods of Willstätter (24). 7 kg. horseradish, washed and cut, are minced and extracted twice with 6 l. of water. The mixture pressed out with a hydraulic press yields 14 l. of extract (P.Z.* 0·65). The extract is saturated with ammonium sulphate; the precipitate is filtered, resuspended in 400 ml. of water and dialysed to remove all salts. About 600 ml. of this fluid is mixed with 2 vol. of 90 % alcohol. The centrifuged precipitate (yield 8 g. dry weight) has a P.Z. of 14–26. The remaining fluid is mixed with a further 1200 ml. of alcohol. The centrifuged precipitate (1·3 g. dry weight) has a P.Z. of 112–300. By continued reprecipitation of the enzyme from aqueous solution with alcohol and by adsorption on and elution from tricalcium phosphate, it is possible to attain a P.Z. of 550. Great losses accompany attempts at further purification. Keilin & Mann (23) report that by employing adsorption on alumina A and precipitation by tannic acid they have raised the activity to P.Z. 1500.

Willstätter & Pollinger (24) observed that their peroxidase pre-

---

* P.Z. or purpurogallin number is expressed as mg. of purpurogallin formed from pyrogallol in 5 min. at 20° per mg. of dry weight of peroxidase preparation. The estimation is carried out in presence of 12·5 mg. of H_2O_2 and 1·25 g. of pyrogallol in 500 ml. of water. P.Z. is the usual unit of peroxidase activity.

parations were always coloured and contained iron. Kuhn *et al.* (25) claimed to have found a close parallelism between the enzymic activity of horseradish peroxidase preparations and the total Fe porphyrin content of these preparations. Keilin, Elliott and Mann (23,26), although unable to confirm quantitatively the results of Kuhn, were able to bring forward new and more reliable evidence supporting the view that the active group of peroxidase is an Fe porphyrin compound.

The absorption spectrum of peroxidase consists of four bands: I, 645 mμ.; II, 538 mμ.; III, 548 mμ. and IV, 498 mμ. On boiling or treating with excess of acid or alkali, these bands disappear, showing that the absorption spectrum belongs to a natural thermolabile protein combined with Fe porphyrin. Keilin & Mann split off the prosthetic group of peroxidase and obtained Fe protoporphyrin. The latter dissolved in pyridine and treated with $Na_2S_2O_4$ gave a typical pyridine-protoferrohaemochromogen. The Fe porphyrin nucleus in peroxidase is therefore identical with that of haemoglobin.

The spectrum of peroxidase conforms to that of Fe porphyrin compounds in the ferric state such as methaemoglobin (23). On addition of $Na_2S_2O_4$ peroxidase undergoes reduction, turning distinctly red, and the four-banded spectrum is replaced by two bands at 594·5 and 558 mμ. This reduced divalent peroxidase forms a reversible compound with CO. Oxidized peroxidase combines reversibly with NaF, KCN, H_2S and NO. These reagents are known to inhibit peroxidase activity.

On addition of H_2O_2 to a strong solution of peroxidase (23) the colour of the solution turns red and the four-banded absorption spectrum is replaced by two bands at 561 and 530·5 mμ. The amount of H_2O_2 necessary to form this compound corresponds to one molecule of H_2O_2 per atom of porphyrin iron present in the preparation. This indicates clearly the formation of a definite enzyme-substrate complex consisting of one molecule of H_2O_2 and one molecule of peroxidase. On further addition of H_2O_2 to a solution of enzyme which already shows a distinct absorption spectrum of the H_2O_2 peroxidase compound the absorption bands of the latter are replaced by two very strong bands lying at 583 and 545·5 mμ. The new bands are most intense when the amount of H_2O_2 added to the solution corresponds to about 100 molecules of peroxide per iron atom of peroxidase.

Peroxidase therefore forms two distinct[23] enzyme-substrate compounds with H_2O_2. The first is formed when peroxidase and H_2O_2 are in more or less equivalent concentrations. The second is formed only in presence of a large excess of H_2O_2. Both these compounds are unstable. In presence of a substance undergoing peroxidatic oxidation, such as hydroquinone or ascorbic acid, the H_2O_2 peroxidase compounds are rapidly decomposed and the original peroxidase is regenerated. It appears therefore that in the catalytic functioning of peroxidase there are two stages: (1) the formation of an enzyme-substrate complex between ferric peroxidase and H_2O_2, and (2) the decomposition of this compound in presence of a suitable substrate, whereby the substrate becomes oxidized by peroxide while ferric peroxidase is regenerated. Unlike catalase, peroxidase remains in the ferric form during the whole cycle of the catalysis. The spectroscopic evidence does not support the view that the substances oxidized are "activated" by the enzyme.

Peroxidase can be obtained from cow's milk by the method of Elliott[19]. A purified preparation from this source shows the same absorption spectrum and the same properties as peroxidase from horseradish roots. Furthermore, the same derivatives are formed by treating the enzyme solution with $Na_2S_2O_4$, CO, KCN, NaF and H_2O_2[23].

Elliott[19] has compared the oxidations catalysed by horseradish and milk peroxidases. Nitrite and tryptophan are not oxidized by H_2O_2 with the plant enzyme but are attacked in presence of the milk enzyme. Dixon[48] has found that persulphate can replace H_2O_2 as oxidant in the case of the milk enzyme but not in the case of the horseradish enzyme. In all other respects the two enzymes behave similarly.

Cytochromes

The cells of all aerobic organisms contain a group of Fe porphyrin proteins (other than haemoglobin, catalase and peroxidase) referred to collectively as the cytochromes[22, 27, 28, 30]. The discovery of these compounds was made by MacMunn in 1886.

It was not, however, until 1925 that the physiological significance of the cytochromes was brought to light by Keilin. The cytochromes are characterized by the ability to undergo reversible oxidation and reduction *in vivo* and by the sharp absorption bands of the reduced compounds contrasted to the diffuse absorption of the oxidized compounds. In the majority of aerobic cells, three distinct cytochromes are observed, and they are referred to as cytochromes *a*, *b* and *c*. Some cells may contain only one cytochrome component, others two components, whereas certain bacteria, e.g. *Bact. coli* (30), contain a cytochrome-like substance which is not identical with any of the three usual components. The identity of the cytochromes from diverse sources is indicated on the basis of the positions of the absorption bands of the reduced compounds.

Cytochrome *c* is the most stable component and is the only one which has been properly extracted and studied *in vitro*. The following are the details of the method of Keilin & Hartree (30) for its preparation from heart muscle of ox or horse. Minced heart is mixed with an equal volume of 0·15 N trichloracetic acid. The centrifuged clear extract shows strong absorption bands of reduced cytochrome together with weak bands of oxyhaemoglobin. Ammonium sulphate is added to the extract (10 g. per 100 ml.) and the precipitate of inactive protein and haemoglobin is discarded. On further addition of trichloracetic acid, oxidized cytochrome *c* is precipitated. The red solid is suspended in water and dialysed against 1 % NaCl. The yield from one ox heart weighing 1100 g. is 30 ml. of a clear dark red solution containing 0·182 g. of cytochrome *c*.

The iron content at the final stage of the above preparation is 0·34 %. Theorell & Åkesson (42) claim to have obtained further purification by electrophoresis first in alkaline solution (pH 10·5) and then in neutral solution (pH 7·3). The iron content of the cataphoretically homogeneous preparation is 0·43 %, corresponding to a molecular weight of 13,000.

Treatment of cytochrome *c* with glacial acetic acid and HBr splits off the iron and liberates haematoporphyrin, which differs from protoporphyrin in having two oxyethyl instead of two vinyl side chains (Hill & Keilin (32), Theorell (31)). Under milder conditions of splitting, a porphyrin is obtained which is not identical either with proto- or haematoporphyrin. Both the nature of the native porphyrin and the mode of linkage of the Fe porphyrin to the protein are obscure at the present time though many attempts have been made to settle these points (cf. Zeile & Meyer(49)).

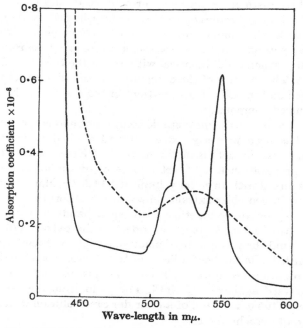

Fig. 5. The visible absorption spectra of oxidized and reduced cytochrome c. The continuous line represents the curve of reduced cytochrome; the dotted line represents that of oxidized cytochrome. (Dixon, Hill & Keilin [1931]. *Proc. Roy. Soc.* B, **109**, 29.)

Cytochrome c, extracted either from heart muscle or from yeast(30,32,33) shows two strong bands in the visible portion of the spectrum when in the ferro form, and only diffuse absorption in the ferri form (cf. Fig. 5). Theorell & Åkesson(42) find that ferricytochrome c exists in four different forms depending upon the pH. The different forms with different absorption spectra change reversibly into one another by changing the pH. The three dissociation constants (pK') are 2·5, 9·5 and 12·7 respectively.

When cytochrome c is reduced by hyposulphite, cysteine, etc., the sole chemical change is the reduction of the trivalent iron in the porphyrin to the divalent state (33,34,35). Since there is one Fe porphyrin for each molecule of cytochrome c only one equivalent of hydrogen or one electronic equivalent is involved in the reduction of each molecule. The oxidation of reduced cytochrome by H_2O_2, ferricyanide, etc. converts iron from the ferrous to the ferric state. The protein which is attached to the Fe porphyrin plays no role in the process of oxidation-reduction. However, the peculiarities which distinguish cytochrome c from a large number of haemochromogen compounds are referable to the specific protein.

The spectrum bands of reduced cytochrome c persist even when a vigorous stream of oxygen is passed through the neutral solution. In other words cytochrome c does not undergo direct oxidation by molecular oxygen. There must be therefore some catalytic mechanism for the oxidation of cytochrome c in vivo. Below pH 4 and above pH 12 reduced cytochrome c is autooxidizable. But in the physiological range of hydrogen-ion concentration it is quite certain that uncatalysed oxidation by molecular oxygen does not occur to any appreciable extent.

Keilin (22,28,29), in his classic researches on cytochrome, has shown that all animal cells as well as aerobic yeasts and bacteria contain an enzyme which catalyses the rapid oxidation of reduced cytochrome. This enzyme is known as the cytochrome oxidase. It is easily prepared from heart muscle, and when added to a solution of reduced cytochrome c in contact with oxygen causes the immediate discharge of the characteristic bands. Reduced cytochrome c is converted to the ferric state, whereas oxygen is reduced to water. The chemical mechanism of this oxidation is not known. All that can be stated is merely that there is a change from ferrous to ferric and that there is a disappearance of oxygen without the simultaneous production of H_2O_2.

Thus far the discussion has been limited to cytochrome c.

Comparatively little is yet known about components a and b. Apart from the fact that components a and b are extremely fragile they are found in association with insoluble particles from which they have not as yet been removed. Our present knowledge of the properties of a and b is mainly derived from the spectroscopic study of a preparation made from heart muscle by the following method (37,38). The muscle is minced and washed exhaustively for 24 hr. until the wash fluid is colourless. The mince is then ground with sand and phosphate buffer to a fine paste and the mixture is centrifuged. The supernatant fluid is brought to pH 4·6 with acetic acid and the precipitate is resuspended in phosphate buffer.

The spectroscopic examination of the cytochromes in the heart preparation is best carried out with a small dispersion spectroscope or microspectroscope replacing the ocular eyepiece of a microscope. Precision spectroscopes with large dispersions are valueless for this type of work. The heart preparation contains the three cytochrome components a, b and c in the oxidized forms. Hence only the faint bands of the oxidized forms are visible. On addition of hyposulphite or succinate the sharp bands of the reduced cytochromes appear at once. The principal absorption bands of the cytochrome components are as follows:

Visible absorption bands (in mμ.) of cytochrome components in heart-muscle preparation (40,44)

	α	β	γ
Cytochrome a	605	?	452
Cytochrome a_3	600	?	448
Cytochrome b	564	530	432
Cytochrome c	550	521	415

Each component has three absorption bands α, β and γ. The notation $c\gamma$, for example, means the γ band of component c. The position of the β band, if any, of component a is not known. It is noteworthy that there is a component a_3 in addition to a. The relation of a_3 to the cytochrome oxidase will be considered later in this chapter.

Component a (40) is an Fe porphyrin protein which is attached to insoluble particles. Above 52° it undergoes rapid destruction.

It does not combine with KCN, CO or other respiratory inhibitors. The oxidized or ferric form is rapidly reducible by the
succinic enzyme system; the ferrous or reduced form is non-
autoxidizable in air.

Component b(40) is also a thermolabile and insoluble Fe
porphyrin protein. The reduced form is to some extent autoxidizable in air in presence of cyanide, a reagent which completely
prevents the oxidation of ferro components a and c. Urethane
inhibits the catalytic reduction of ferric components a and c, and
the oxidation of ferro component b. The mechanism of this inhibition is obscure. Respiratory inhibitors such as KCN, CO, etc. do
not combine with component b either in the ferro or ferric forms.

A suspension of yeast examined with the microspectroscope
shows the bands of the reduced cytochromes. On shaking
vigorously with air the bands disappear. This process can be repeated many times with the same suspension. This shows clearly
that the cytochromes undergo a reversible cycle of oxidation and
reduction. If the yeast suspension is treated with reagents which
specifically interfere with the reducing systems, then the cytochromes exist mainly in the oxidized state. Conversely, when the
cytochrome oxidase is poisoned by cyanide or H_2S, the cytochromes are found mainly or even entirely in the reduced state.
In later chapters the nature of the systems which are concerned
in the reduction of the cytochromes will be considered in some
detail.

There is a rough parallelism between the respiratory activity
of aerobic cells and the concentration of both cytochrome and
cytochrome oxidase. The higher the concentration of cytochrome
and oxidase, the greater the respiratory activity. This interesting
correlation led Keilin to the view that the respiratory process
proceeds mainly through the cytochromes. There is evidence
strongly in favour of this view. The respiration of many aerobic
cells is largely inhibited by cyanide—a reagent which is highly
poisonous to the cytochrome oxidase although without effect on
many other oxidation enzymes. Recently Haas(36), using an
elegant spectroscopic method, determined the rates with which
the cytochromes are reduced and oxidized in living yeast, and on
the basis of these data he was able to calculate the percentage of

the total respiration which proceeds through the cytochromes. His calculation indicated that practically the entire respiration was linked up with the cytochromes. That is to say, for each atom of oxygen absorbed by yeast two molecules of cytochrome are reduced; and for each molecule of hydrogen* removed from the substrates by oxidation, two molecules of cytochrome are reduced.

Cytochrome oxidase

Reduced cytochromes a and c are not autoxidizable in air at physiological hydrogen-ion concentrations. The oxidation by molecular oxygen is catalysed by the cytochrome oxidase. A highly active preparation of the enzyme is easily made from heart muscle (37,38) of pig, ox, or sheep by the method described above for the preparation of cytochromes a and b. The enzyme is associated with insoluble particles which are easily sedimented by centrifugation. No success has yet attended efforts to purify the enzyme. It is removed by filtration through charcoal, kaolin and kieselguhr. Acetone treatment or drying destroys the activity of the enzyme.

The cytochrome oxidase specifically catalyses the oxidation of the ferro cytochromes by molecular oxygen (38). Haemoglobin and other ferrous porphyrin compounds are not oxidized in presence of the enzyme. For a long time the cytochrome oxidase was attributed with the property of catalysing the oxidation of p-phenylenediamine. The error arose from the fact that preparations of the enzyme from heart muscle contain small amounts of the cytochromes which can oxidize p-phenylenediamine directly. Cytochrome thus acts as a catalyst for the oxidation of p-phenylenediamine by undergoing a cycle of reduction by the diamine and oxidation by molecular oxygen. Preparations of the cytochrome oxidase which contain no cytochrome have no influence

---

* It may be difficult at first glance to understand how hydrogen can be transferred to cytochrome which suffers only a valency change in the course of reduction. In the reduction of ferric compounds by hydrogen, the complete reaction may be formulated as follows:

$$Fe^{+++} + H \rightarrow Fe^{++} + H^+$$

An electron is shifted from atomic hydrogen to the ferric ion and a hydrogen ion is thereby produced.

on the aerobic oxidation of p-phenylenediamine. The system cytochrome oxidase—cytochrome c can also catalyse the oxidation of cysteine, catechol, adrenaline and ascorbic acid (cf. Fig. 6). Here again the reaction between these substrates and molecular oxygen proceeds via the intermediation of cytochrome c. The function of the oxidase is to catalyse the oxidation of ferro

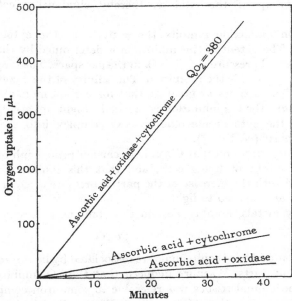

Fig. 6. The catalytic action of cytochrome c on the oxidation of ascorbic acid in presence of the cytochrome oxidase and oxygen. (Keilin & Hartree [1938]. *Proc. Roy. Soc.* B, **124**, 397.)

cytochrome c by oxygen. The reduction of ferri cytochrome c by the various substrates does not require any catalyst.

The rate of oxidation of the ferro cytochromes by molecular oxygen in presence of the enzyme is practically independent of the partial pressure of oxygen. It is significant that the respiration of yeast and other micro-organisms is also independent of the oxygen tension. This and many other lines of evidence point to the cytochrome oxidase as the essential catalyst in aerobic events.

One of the most striking characteristics of the cytochrome oxidase is the extraordinary sensitivity to HCN, H_2S and azide. $M/1000$ concentrations of any of these substances completely inhibit the activity of the oxidase. There is good evidence that these inhibitors do not combine with cytochromes a, b and c. The only alternative possibility is that combination takes place with the active group of the oxidase, thereby eliminating catalytic activity.

Carbon monoxide inhibits the activity of the cytochrome oxidase. The extent of this inhibition is determined by the ratio of the partial pressure of CO to O_2 in the gas space. The larger the ratio the greater is the inhibition. The affinity of the enzyme for oxygen is five times as great as that for carbon monoxide. In strong light the CO inhibition is reduced. Light apparently dissociates the cytochrome oxidase—CO complex into the component parts (cf. Fig. 7).

Warburg[41] found that CO in high concentrations inhibits the oxygen uptake of living cells, and that this inhibition is diminished with the increase of the partial pressure of O_2 or when the cells are exposed to light.

Heavy metals, notably Fe and Cu, combine reversibly with CO, e.g.—

$$Fe^{++} + CO \rightleftharpoons FeCO$$

These iron-carbonyl compounds are dissociated by light into their component parts. Warburg deduced that the CO inhibition of yeast and animal tissues was referable to some iron compound which he called the "Respiratory Enzyme". This enzyme normally undergoes a cycle of valency change from the ferrous to the ferric condition. In presence of CO the oxidation of the ferrous enzyme by molecular oxygen is inhibited by the formation of a compound between CO and the ferrous enzyme. There is thus a competition between CO and O_2 for the ferrous enzyme. The ratio $[Fe^{++}]/[FeCO]$ depends on the ratio of the partial pressures P_{CO}/P_{O_2}:

$$\frac{[Fe^{++}]}{[FeCO]} \times \frac{P_{CO}}{P_{O_2}} = K.$$

Since the affinity of the enzyme for O_2 is 5–9 times as great as for CO it is necessary to have a large excess of CO to obtain effective inhibition. The reversal of the CO inhibition by light was studied quantitatively by Warburg, notably in yeast and acetic-acid bacteria. The efficiency of light of different wave-lengths was found to follow a curve which strongly resembled the light-absorption curve of an Fe porphyrin compound (41). Hence Warburg concluded that the "Respiratory Enzyme" is in fact an Fe porphyrin compound. The assumption on which this deduction rests is that the wave-lengths at which the respiratory enzyme—CO compound absorbs strongest should be the most efficient for photo-decomposition, and conversely that those wave-lengths of light which are weakly absorbed should be less effective. Experiments on model systems have confirmed the quantitative validity of this assumption.

It is important to note that Warburg had no detailed information as to the function of his respiratory enzyme apart from knowing that it played a fundamental role in oxidation processes. The identity of Warburg's enzyme with the cytochrome oxidase was established only after the latter was isolated and studied in cell-free preparations. Usually the discovery of the function of an enzyme precedes the discovery of the chemical nature of the prosthetic group. But in this case the time relations were reversed. The chemical nature of the prosthetic group of the respiratory enzyme was discovered before its specific function was known.

Keilin & Hartree (40) have found that component a of cytochrome in heart muscle is made up of two distinct components a and a_3. The properties of a_3 are quite unlike those of the other three cytochrome components. It combines with KCN either in the ferro (a_3^{++}) or ferric (a_3^{+++}) form. The a_3^{++} compound with cyanide is rapidly autoxidizable in air to the a_3^{+++} state. The a_3^{+++} compound with cyanide is not easily reducible by the succinic enzyme system. Component a_3^{+++} also combines with H_2S and azide. a_3^{++} combines with CO to form a compound whose two principal bands in the visible spectrum occupy the same positions as the corresponding bands in the photochemical spectrum of the CO inhibition of respiration.

34   FE PORPHYRIN PROTEIN ENZYMES

Absorption bands of CO *compound* (*in* mμ.)

(Kubowitz & Haas [1932]. *Biochem. Z.* 255, 247;
Keilin & Hartree [1939]. *Proc. Roy. Soc.* B, 127, 167.)

	α	β	γ
"Respiratory Enzyme" (yeast)	590	540	430
Cytochrome a_3	590	Masked	430

In many respects a_3 shows the characteristic properties of the cytochrome oxidase, e.g. reaction with CO, KCN, H_2S and azide. It is also destroyed by the same agents and conditions as the cytochrome oxidase, e.g. by fat solvents, drying, temperatures higher than 52°, acid or alkali. But the correspondence in properties is not complete. Cytochrome oxidase by definition catalyses the oxidation of the ferro cytochromes. a_3 in the oxidized form appears incapable of oxidizing ferro cytochrome *c* under anaerobic conditions. Furthermore, the CO inhibition of the cytochrome oxidase is light sensitive, whereas the CO compound of a_3^{++} is apparently insensitive to light. Suitable explanations can be found for these discrepancies, but for the moment it would be premature to identify a_3 with the cytochrome oxidase.

There is an interesting relation between components *a* and a_3. The intensities of the *a* and a_3 bands always run parallel. The indications are that *a* and a_3 have the same porphyrin nucleus, since they both yield the same haemochromogen. The porphyrin nucleus of these two components is probably similar to that of chlorocruorin. Keilin & Hartree suggest that *a* and a_3 may be interconvertible.

Many aerobic bacteria are lacking in component *a* and hence a_3. The respiration of such cells is inhibited by KCN and CO, although the CO inhibition is not light sensitive. The absence of a_3 in such cells is hard to fit in with the presence of a typical cyanide-sensitive cytochrome oxidase.

REFERENCES

(1) Laidlaw [1904]. *J. Physiol.* 31, 464.
Milroy [1909]. *J. Physiol.* 38, 384.
Hill [1925]. *Biochem. J.* 19, 341.
(2) Hill & Holden [1926]. *Biochem. J.* 20, 1326; [1927]. *Ibid.* 21, 625.
(3) Anson & Mirsky [1925]. *J. Physiol.* 60, 50.

(4) Keilin [1926]. *Proc. Roy. Soc. B*, 100, 124.
(5) Coryell, Stitt & Pauling [1937]. *J. Amer. Chem. Soc.* 59, 633.
(6) Keilin [1933]. *Proc. Roy. Soc. B*, 113, 393.
(7) Keilin [1936]. *Proc. Roy. Soc. B*, 121, 165.
(8) Austin & Drabkin [1935]. *J. biol. Chem.* 112, 67.
(9) Hill [1926]. *Proc. Roy. Soc. B*, 100, 419.
(10) Zeile & Hellström [1930]. *Hoppe-Seyl. Z.* 192, 171.
(11) Stern [1936]. *J. biol. Chem.* 112, 661.
(12) Sumner & Dounce [1939]. *J. biol. Chem.* 127, 439.
(13) Lemberg [1937]. *Perspectives in Biochemistry*, Cambridge University Press, pp. 137–49.
(14) Keilin & Hartree [1936]. *Proc. Roy. Soc. B*, 121, 173.
(15) Keilin & Hartree [1938]. *Proc. Roy. Soc. B*, 124, 397.
(16) Stern [1936]. *J. biol. Chem.* 114, 473.
(17) Keilin & Hartree [1936]. *Proc. Roy. Soc. B*, 119, 114.
(18) Euler [1934]. *Chemie d. Enzyme III.* Munchen.
(19) Elliott [1932]. *Biochem. J.* 26, 10; 26, 1281.
(20) Willstätter [1928]. *Untersuchungen über Enzyme*, Berlin.
(21) Bancroft & Elliott [1934]. *Biochem. J.* 28, 1911.
(22) Keilin [1929]. *Proc. Roy. Soc. B*, 104, 206.
(23) Keilin & Mann [1937]. *Proc. Roy. Soc. B*, 122, 119.
(24) Willstätter & Pollinger [1932]. *Ann. Phys., Lpz.*, 430, 269.
(25) Kuhn, Hand & Florkin [1931]. *Hoppe-Seyl. Z.* 201, 255.
(26) Elliott & Keilin [1934]. *Proc. Roy. Soc. B*, 114, 210.
(27) MacMunn [1887]. *J. Physiol.* 8, 57; [1889]. *Hoppe-Seyl. Z.* 13, 497.
(28) Keilin [1925]. *Proc. Roy. Soc. B*, 98, 312.
(29) Keilin [1930]. *Proc. Roy. Soc. B*, 106, 418; [1933]. *Ergebn. Enzymforsch.* 2, 239.
(30) Keilin & Hartree [1937]. *Proc. Roy. Soc. B*, 122, 298.
(31) Theorell [1938]. *Biochem. Z.* 298, 242; [1939]. *Enzymologia*, 6, 88.
(32) Hill & Keilin [1930]. *Proc. Roy. Soc. B*, 107, 286.
(33) Dixon, Hill & Keilin [1931]. *Proc. Roy. Soc. B*, 109, 209.
(34) Hill & Keilin [1933]. *Proc. Roy. Soc. B*, 114, 104.
(35) Theorell [1935]. *Biochem. Z.* 279, 463.
(36) Haas [1934]. *Naturwissenschaften*, 22, 207.
(37) Ogston & Green [1935]. *Biochem. J.* 29, 1983.
(38) Keilin & Hartree [1938]. *Proc. Roy. Soc. B*, 125, 171.
(39) Agner [1938]. *Biochem. J.* 32, 1702.
(40) Keilin & Hartree [1939]. *Proc. Roy. Soc. B*, 127, 167.
(41) Warburg [1932]. *Z. angew. Chem.* 45, 1.
(42) Theorell & Åkesson [1939]. *Science*, 90, 67.
(43) Lemberg, Norrie & Legge [1939]. *Nature, Lond.*, 144, 551.
(44) Warburg & Negelein [1931]. *Biochem. Z.* 233, 486; 238, 135.
(45) Lemberg [1935]. *Biochem. J.* 29, 1322.
(46) Fischer & Klarer [1926]. *Ann. Phys., Lpz.*, 448, 178.
(47) Willstätter & Fischer [1913]. *Hoppe-Seyl. Z.* 87, 423.
(48) Dixon [1934]. *Biochem. J.* 28, 2061.
(49) Zeile & Meyer [1939]. *Hoppe-Seyl. Z.* 262, 178.

CHAPTER III

Pyridinoprotein enzymes

Harden & Young[1] in 1906 discovered that yeast juice after dialysis lost the power of fermenting glucose. This power was restored by addition of the dialysate. The term cozymase was applied to the crystalloidal, thermostable substance the presence of which was essential to the process of fermentation. Following this pioneer discovery Euler & Myrbäck made important contributions to our knowledge of its chemical constitution. The back of the cozymase problem was finally broken when Warburg and his institute focused their attention on the isolation of an oxidation coenzyme which they found in blood corpuscles.

Warburg & Christian[2] observed that mammalian red blood corpuscles contained an enzyme which catalysed the oxidation of aldohexosemonophosphate (glucose-6-phosphate). This enzyme required the collaboration of a thermostable substance of low molecular weight which they called the coenzyme. From 250 l. of horse blood they succeeded in obtaining *ca.* 1 g. of a highly purified product, 0·001 mg. of which could transfer 1 μl. of O_2 per min. in presence of the hexosemonophosphoric enzyme system. The Q_{O_2} (μl. O_2 per hr. per mg. dry weight) of the pure coenzyme is about 60,000.

The method of preparation of the coenzyme from red blood corpuscles is too involved to describe in detail here. The following are the essential steps of the method[3]. Corpuscles washed free of serum are laked with water. Haemoglobin is removed by addition of an alcohol-chloroform-ether mixture. The haemoglobin-free filtrate is mixed with a large excess of alcohol and ether, and the coenzyme is precipitated. The barium salt of the coenzyme is soluble in water at pH 8·4, whereas the barium salts of the principal contaminants (adenylic acid, diphosphoglyceric acid and other phosphoric compounds) are insoluble. After fractionation of the barium salts the insoluble mercury salt of the coenzyme is formed and decomposed with H_2S. To remove the last traces of adenylic acid advantage is taken of the fact that the coenzyme is soluble in methyl alcohol $N/10$ with respect to HCl, whereas adenylic acid is insoluble. Final purification involves fractional alcohol precipitation of the lead salt of the coenzyme.

The coenzyme[3] is composed of one molecule of adenine, one of nicotinicamide, three of phosphoric acid and two of pentose. The coenzyme as isolated is a hydrate with six molecules of water. The analytical data agree with the formula $C_{21}H_{28}O_{17}N_7P_3$. Two of the phosphoric acid groups are titratable, whereas the third is substituted. The following is a provisional formula of its constitution, suggested by Euler et al.[4].

In presence of the specific enzyme, hexosemonophosphate is oxidized to phosphohexonate, while the coenzyme is simultaneously reduced[3]. One molecule of hydrogen becomes thereby transferred from the substrate to the coenzyme. This reduction of the coenzyme can also be brought about by hyposulphite. The coenzyme when reduced either by the enzyme system or by hyposulphite can be reoxidized to the original compound by flavoprotein. This reversible oxidation and reduction is the key to the catalytic function of the coenzyme. When the coenzyme is reduced by Pt and H_2 in a slightly alkaline solution, three molecules of hydrogen are taken up per molecule of coenzyme. The

reduction product thus formed cannot be oxidized by flavoprotein. That is to say, reduction with nascent hydrogen is an
over-reduction and is not reversible. The catalytic activity of the
coenzyme is correspondingly destroyed.

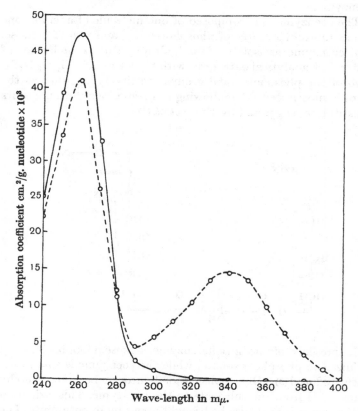

Fig. 7. Absorption spectrum of oxidized (o——o) and reduced (o- - -o)
coenzyme II. (Warburg & Christian [1936]. *Biochem. Z.* **287**, 20.)

The coenzyme has no absorption bands in the visible region of
the spectrum, but shows a broad band at 260 mμ. in the u.v.(3)
which can easily be accounted for as the summation of the individual absorptions of both the adenine and nicotinicamide

moieties of the molecule. On reversible reduction this band remains unchanged and a new band at 345 mμ. appears (cf. Fig. 7). On irreversible reduction of the coenzyme the band at 260 mμ. fades and the absorption that remains is due entirely to the adenine moiety. From this evidence Warburg[3] concluded that irreversible reduction transforms the pyridine ring of nicotinicamide into piperidine. The latter does not absorb appreciably at 260 mμ. The fading of the 260 band is thereby explained. If the coenzyme is first reduced enzymically and then treated with Pt and H$_2$, two molecules instead of three of hydrogen are taken up in the second stage. This fact shows clearly that in the biological reduction of the coenzyme it is only the pyridine ring which takes up a molecule of hydrogen.

Warburg's interpretation of the spectroscopic data received confirmation from a similar study of the methiodide of nicotinicamide[5]. This substance is reduced in weakly alkaline solution to a dihydro-derivative:

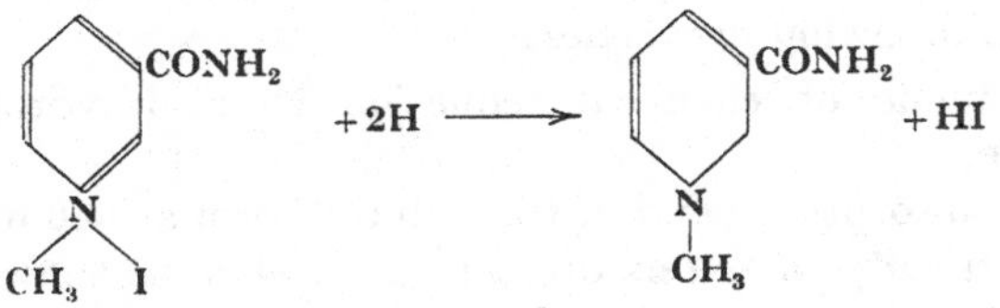

The partially reduced form can be reoxidized by various oxidizing agents to the original pyridine compound. On reduction to o-dihydropyridine, a broad absorption band appears at 360 mμ., similar in shape to the 345 band of the coenzyme.

The isolation of nicotinicamide from the coenzyme of the hexosemonophosphoric system (now known as coenzyme II or as triphosphopyridinenucleotide) led Euler et al.[7] to reinvestigate the problem of the chemical constitution of cozymase (now known as coenzyme I in virtue of its being the first to be discovered). Both Euler and Warburg independently found that coenzyme I contains all the components of coenzyme II in the same proportions except for the phosphoric acid. Coenzyme I contains only two molecules of phosphoric acid in contrast to the three of coenzyme II. A provisional formula (4) of its constitution is given below:

On reduction:

In common with coenzyme II the diphosphopyridinenucleotide shows the following properties:

(1) Enzymic or chemical reduction to a dihydropyridine derivative;

(2) An absorption band of the reduced form at 340 mμ.;

(3) Instability of the oxidized form in alkaline solution, and of the reduced form in acid solution;

(4) Non-autoxidizability of the reduced form;

(5) Oxidation of the reduced form by specific flavoproteins.

Coenzyme I is widely distributed in the animal, plant and bacterial kingdoms. In fact it may be considered as part of the essential equipment of all cells. Yeast is extremely rich in coenzyme I and offers the most suitable source of preparation. The principles of the isolation are essentially the same as for that of coenzyme II, except that the barium and lead salts of coenzyme I are relatively more soluble. The complete separation of coenzyme I from II offers considerable difficulties, owing to the close similarity of the respective chemical properties.

Despite the almost complete chemical correspondence between coenzymes I and II their biological properties are not similar.

Triphosphopyridinenucleotide acts as coenzyme for the hexose-monophosphoric, citric and phosphohexonic enzymes. Diphospho-pyridinenucleotide is inactive in these two systems but functions catalytically· in the lactic, malic, β-hydroxybutyric, aldehyde, triosephosphoric, α-glycerophosphoric and glyceraldehyde enzyme systems. The glucose and glutamic enzymes of liver are alleged to be capable of collaborating with either coenzyme I or II. With this exception enzymes show complete specificity for one of the two pyridine nucleotides.

Attempts have been made to replace coenzymes I and II with synthetic pyridine derivatives. Thus far no case has been recorded of a synthetic substance showing coenzyme activity. Schlenk et al.(8) have treated coenzyme I with nitrous acid and found that "desaminocoenzyme" can act catalytically although less effectively than the parent compound.

The class of pyridinoproteins may be subdivided into (1) di-phosphopyridinoproteins and (2) triphosphopyridinoproteins. The enzymes of the first group require the presence of coenzyme I, whereas those of the second group require the presence of coenzyme II. The glucose and glutamic enzymes of liver are exceptional in satisfying the requirements of both groups.

In these two classes of enzymes we have striking examples of a family of oxidation enzymes each member of which is highly specific for the reaction catalysed, though all depend upon the same functional group for their activity. The protein part of the enzyme is not the same for any two members of the family. That is not surprising in view of the extraordinary specificity shown by the various enzymes.

DIPHOSPHOPYRIDINOPROTEIN ENZYMES

Equilibrium between substrate and coenzyme

Let us consider the details of the oxidation of ethyl alcohol catalysed by the alcohol enzyme of yeast. The principles which apply in this case can be extended to all the other pyridino-proteins. A mixture of alcohol and the alcohol enzyme does not react. On addition of coenzyme I a reaction takes place whereby alcohol is oxidized to aldehyde and an equivalent amount of coenzyme I is reduced to the dihydropyridine form. The course of the reaction is easily followed spectrophotometrically (9, 10). The reduced coenzyme shows an absorption band at 340 mμ., whereas the oxidized coenzyme does not absorb in that region of the spectrum. Employing a monochromatic source of light and a photoelectric cell for recording the intensity of the light after passing through the solution, the decrease in the light intensity at 340 mμ. can be equated with the extent of reduction of the coenzyme. Fig. 8 shows a record of such an experiment.

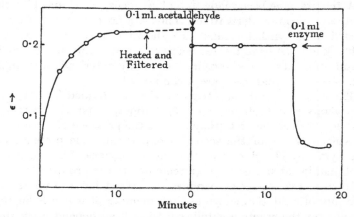

Fig. 8. The reduction of coenzyme I by alcohol, and the oxidation of reduced coenzyme I by acetaldehyde in presence of the yeast alcohol enzyme. The initial reaction mixture contained enzyme, coenzyme I and alcohol. (Euler, Adler & Hellström [1936]. *Hoppe-Seyl. Z.* 241, 239.)

The extinction coefficient at first rapidly increases and then slowly reaches a plateau. The theoretical extinction coefficient for complete reduction of the coenzyme is much higher than the observed asymptotic value. The discrepancy suggests that there is an equilibrium between ethyl alcohol and coenzyme on the one hand and acetaldehyde and reduced coenzyme I on the other. If such were the case the addition of excess aldehyde should reverse the reaction and the coenzyme should almost completely revert to the oxidized form. This prediction is realized experimentally (cf. Fig. 8). It is significant that the reaction between

aldehyde and reduced coenzyme takes place only in the presence
of the specific enzyme.

The above observations may be summarized by the following
equation:

$$\text{alcohol} + \text{coenzyme I} \xrightleftharpoons{\text{enzyme}} \text{aldehyde} + \text{reduced coenzyme I}$$

The equilibrium constant KK for the reaction is defined by the
equation:

$$\frac{[\text{coenzyme I}]\,[\text{alcohol}]}{[\text{reduced coenzyme I}]\,[\text{aldehyde}]} = KK,$$

brackets indicating concentration in mols. per litre. At 20° C.
and pH 7·9, $KK = 1·09 \times 10^3$. For half-reduction of the coenzyme
under these conditions the ratio of $\dfrac{\text{alcohol}}{\text{aldehyde}}$ concentrations must

be $\dfrac{1090}{1}$ (11, 12). At pH 7·0 and 38°, $KK = 2·14 \times 10^4$. The equilibrium is greatly in favour of the oxidation of reduced coenzyme
I by acetaldehyde.

The simple experimental facts give no clue as to the relation
between the coenzyme and the enzyme. Warburg believes that
the coenzyme is the prosthetic group of the specific protein for
alcohol oxidation and that the compound of enzyme and coenzyme—unlike haemoglobin, in which prosthetic group and
globin are rigidly bound—dissociates almost completely into the
constituent parts. Thus according to this view(11) the alcohol
enzyme, i.e. the protein, has no catalytic properties when un-
combined with the coenzymes. In virtue of the dissociation of
the enzyme-coenzyme compound it is possible for one molecule
of the protein to bring about the reduction of several thousand
molecules of coenzyme per minute. The postulated cyclical process may be represented as follows:

protein + coenzyme I $\rightleftharpoons$ protein—coenzyme I

protein—coenzyme I + alcohol $\rightleftharpoons$
    protein—reduced coenzyme I + aldehyde

protein—reduced coenzyme I $\rightleftharpoons$ protein + reduced coenzyme I

As yet there is no conclusive proof of this interpretation and the
reader may regard the coenzyme either as the prosthetic group

or as a highly specific hydrogen acceptor. In the latter case, the protein would be the catalytic unit with the coenzyme in the role of a specific oxidizing agent (cf. Dixon & Zerfas[37]).

Under aerobic conditions yeast oxidizes alcohol to acetaldehyde and eventually to CO_2 and H_2O. Under anaerobic conditions alcohol accumulates. This must mean that the cell can regulate the direction of the equilibrium between the alcohol-aldehyde and coenzyme-reduced coenzyme systems. In order to oxidize a large amount of alcohol in the presence of catalytic amounts of enzyme and coenzyme two conditions must be fulfilled: (1) aldehyde must be removed as quickly as formed and (2) there must be some mechanism for the reoxidation of reduced coenzyme. Yeast contains an enzyme which catalyses the aerobic oxidation of aldehydes and thus can remove aldehyde oxidatively. Furthermore, there are agents in yeast such as flavoprotein which can oxidize reduced coenzyme I and thus make it available for further oxidation of alcohol to aldehyde. In the next chapter we shall consider in detail the mechanisms for the aerobic oxidation of reduced coenzyme.

Coenzyme-linked reactions

Heart muscle contains an enzyme[13] which catalyses the oxidation of β-hydroxybutyric acid by coenzyme I:

$$\beta\text{-hydroxybutyrate} + \text{coenzyme I} \rightleftharpoons$$
$$\text{acetoacetate} + \text{reduced coenzyme I}$$

Here again the reaction reaches an equilibrium. The equilibrium constant KK at pH 7·0 and 38° is about 5000 times smaller than that for the alcohol system under the same conditions. Thus to obtain half reduction of the coenzyme the ratio $\dfrac{\text{alcohol}}{\text{aldehyde}}$ must be $ca.\ \dfrac{21,000}{1}$ and the ratio $\dfrac{\beta\text{-hydroxybutyrate}}{\text{acetoacetate}}$ $ca.\ \dfrac{4\cdot5}{1}$.

The accumulation of acetoacetate must be considerable before equilibrium between β-hydroxybutyrate and coenzyme I is reached, whereas a trace of aldehyde is sufficient to stop the oxidation of alcohol.

Suppose the following mixture is prepared: β-hydroxybutyrate and its appropriate enzyme, coenzyme I, and aldehyde and its appropriate enzyme. A reaction will take place whereby β-hydroxybutyrate is oxidized to acetoacetate whilst aldehyde is reduced to alcohol. The reaction is not direct but occurs in two stages:

$$\beta\text{-hydroxybutyrate} + \text{coenzyme I} \xrightarrow{\text{enzyme}} \text{acetoacetate} + \text{reduced coenzyme I}$$

$$\text{reduced coenzyme I} + \text{aldehyde} \xrightarrow{\text{enzyme}} \text{coenzyme I} + \text{alcohol}$$

The coenzyme acts as a link between the two systems by undergoing reduction by the one and oxidation by the other. The reaction will proceed almost to completion before an equilibrium is reached. Reactions of this type are known as coenzyme-linked [14]. The β-hydroxybutyrate system is referred to as the reducing half and the alcohol-aldehyde system as the oxidizing half.

The KK values for the equilibrium with coenzyme I of the lactate-pyruvate [15], malate-oxaloacetate [16], glutamate-ammonium α-ketoglutarate [17], and α-glycerophosphate-triosephosphate systems are shown in Table III. The value for the triosephosphate-phosphoglycerate system has not yet been determined.

Table III. *Equilibrium constants of pyridinoprotein enzyme systems at 30°*

System	KK	pH	Reference
α-Glycerophosphate-triosephosphate	1.7×10^3	8·1	Euler *et al.* (38)
	3.3×10^3	7·6	
	6.3×10^4	6·9	
Lactate-pyruvate	14.1×10^4	7·4	Euler *et al.* (57)
	2×10^4	8·1	
	6.5×10^3	8·7	
Alcohol-aldehyde	16.6×10^4	6·05	Adler & Sreenivasaya (58)
	1.3×10^4	7·25	
	2×10^3	8·0	
Glutamate-ammonium α-ketoglutarate	2.43×10^4	7·01	Euler *et al.* (17)
	5.5×10^3	7·26	
	1.1×10^3	8·0	
Malate-oxaloacetate	2.85×10^4	7·54	Euler *et al.* (16)
	8.8×10^3	7·85	
	2×10^3	8·44	
β-Hydroxybutyrate-acetoacetate	4·5	7·0	Calculated from data of Green *et al.* (13)

Prediction could be made from the KK values whether oxido-reductions between given pairs of enzyme systems would be possible, and how far these reactions would proceed to completion. These possibilities have in many cases been verified experimentally, e.g.

$$\beta\text{-hydroxy-}\atop\text{butyrate} + \begin{cases}\text{oxaloacetate} \\ \text{pyruvate} \\ \text{aldehyde}\end{cases} \xrightarrow[\text{coenzyme}]{\text{enzymes}} \text{acetoacetate} + \begin{cases}\text{malate} \\ \text{lactate} \\ \text{alcohol}\end{cases}$$

The above reactions proceed almost completely to the right before an equilibrium is reached. Systems with approximately the same KK values have less possibility of reaction. Thus the following oxido-reductions have no practical significance:

$$\begin{aligned}\text{lactate} + \text{oxaloacetate} &\to \text{pyruvate} + \text{malate} \\ \text{malate} + \text{pyruvate} &\to \text{oxaloacetate} + \text{lactate} \\ \text{lactate} + \text{aldehyde} &\to \text{pyruvate} + \text{alcohol}\end{aligned}$$

Apparent exceptions are the oxido-reductions between α-glycero-phosphate on the one hand and pyruvate and oxaloacetate on the other. The KK values of the three pyridinoprotein systems concerned are very close to one another and we should expect that the oxido-reductions would not proceed to completion. The likely explanation of the observation that in muscle extracts completion is reached is that the product of the oxidation of α-glycerophosphate, viz. triosephosphate, can be further oxidized to phosphoglycerate. Since the oxido-reductions were carried out with crude enzyme preparations containing both the α-glycero-phosphate and triosephosphoric enzymes the secondary oxidation of triosephosphate was not eliminated and the oxido-reduction proceeded beyond the true equilibrium point.

Equilibrium constants and oxidation-reduction potentials

The equilibrium constant KK for the reaction between a pyridinoprotein enzyme system and the coenzyme can be translated in terms of an oxidation-reduction potential E. The equilibrium constant and the oxidation-reduction potential are two different ways of expressing numerically the equilibrium state, and the value of one can be calculated from the other. The reader is referred to the monographs of Clark[19] and Michaelis[20] for a complete account of the theory and practice of oxidation-reduction potentials. The following are the oxidation-reduction potentials of some of the pyridinoprotein enzyme systems at pH 7·0:

	E_0' (in volts)	$t°$	Reference
Malate-oxaloacetate	−0·169	37	Laki (53)
Lactate-pyruvate	−0·180	35	Barron & Hastings (55) Wurmser & Mayer-Reich (56)
Alcohol-aldehyde	−0·20 (approx.)	30	Barron (54) Adler & Sreenivasaya (58)
β-Hydroxybutyrate-acetoacetate	−0·293	38	Hoff-Jørgensen (59) Green et al. (13)
Coenzyme I—reduced coenzyme I	−0·29 (approx.)	30	Clark (60) Schlenk et al. (8)

Dismutations and mutases

Mammalian liver contains an enzyme, discovered by Parnas, which catalyses the dismutation of aldehydes, e.g.

$$2CH_3CHO + H_2O \rightarrow CH_3CHOH + CH_3COOH$$

One molecule of aldehyde is oxidized to acetate and simultaneously another is reduced to ethyl alcohol. The presence of coenzyme I is necessary for the reaction to take place. Dixon & Lutwak-Mann[21] have called this enzyme the aldehyde mutase. A mutase is defined as an enzyme or enzyme system which catalyses the oxidation of one molecule of its substrate at the expense of the reduction of another molecule. The dismutation of aldehyde takes place in two stages:

aldehyde + coenzyme I → acetate + reduced coenzyme I
reduced coenzyme I + aldehyde → coenzyme I + alcohol

The aldehyde mutase is therefore equivalent to an alcohol and an aldehyde enzyme combined. It is difficult to decide [22] at the moment whether the mutase is in fact a mixture of two enzymes or whether it is a special pyridinoprotein capable of a dual catalytic role. Dismutations are special cases of coenzyme-linked reactions in which the same compound is at once reductant and oxidant for the coenzyme.

The dismutation of triosephosphate is considered to be one of the reactions in muscle glycolysis:

2 triosephosphate → phosphoglycerate + α-glycerophosphate

One molecule of the phosphorylated triose is reduced to the corresponding alcohol whilst another is oxidized to the corresponding sugar acid. Here again coenzyme I is a participant in the reaction. Adler & Hughes(23) and Quastel & Wheatley(47) have brought forth evidence that two pyridinoproteins are concerned in the catalysis, one catalysing the reduction of triosephosphate to α-glycerophosphate by reduced coenzyme I, the other catalysing the oxidation of triosephosphate to phosphoglycerate by oxidized coenzyme I.

Nomenclature

Before proceeding to a consideration of the individual pyridinoproteins a point of nomenclature is in order. It has long been the tradition to call the specific proteins which collaborate with the coenzyme "dehydrogenases". The objection has been raised that the term dehydrogenase should be applied only to the complex of protein and coenzyme, since the protein by itself is not the catalyst. For the purposes of simplicity and uniformity the term dehydrogenase will be reserved in this book for the complete catalytic system necessary for the activation of the substrate in cases where more than one component is involved. But it is to be remembered that current nomenclature is a matter of convention and has no bearing on the validity of whatever theory is tacitly assumed in the terms used.

Lactic and malic enzymes (animal tissues)

Although these proteins are different, their properties and kinetics are so very similar that it would be simplest to consider them together. The method of preparation of the two enzymes is as follows(24). The stripped skeletal muscle of rabbit is minced and mixed with iced water. After 30 min. the brei is filtered through muslin and the filtrate is mixed with 3 vol. cold acetone. The precipitate is filtered, washed with acetone and ether, and pulverized. The powder is rubbed up with water and the suspension is dialysed. The insoluble material is then centrifuged off. The clear and faintly coloured supernatant fluid contains highly active lactic and malic enzymes.

There are two experimental methods of studying the catalytic oxidation of lactate or malate. The direct method involves spectrophotometric measurement of the rate of reduction of the coenzyme by the substrates in presence of their respective enzymes. The indirect method involves the expedient of adding an additional hydrogen acceptor such as methylene blue or oxygen, and by following either colorimetrically or manometrically the rate of disappearance of these acceptors it is possible to obtain a quantitative measure of the rate of oxidation of the substrate by the coenzyme. The reaction between reduced coenzyme and these acceptors is complex and involves other catalysts, which will be considered in the next chapter.

The keto acids formed by oxidation of lactate or malate, viz. pyruvate and oxaloacetate, inhibit[25, 26] almost completely the oxidation of the respective hydroxy acids even when present in very minute concentration (ca. $M/1000$). Strictly speaking, the action of the keto acids is not entirely one of inhibition. There is the additional consideration that the keto acids are forcing the reaction to proceed in the reverse way, i.e. towards oxidation of the coenzyme and not towards reduction. The equilibria may be formulated as follows:

lactate + coenzyme $\rightleftharpoons$ pyruvate + reduced coenzyme

malate + coenzyme $\rightleftharpoons$ oxaloacetate + reduced coenzyme

Therefore in order to study the enzymic oxidation of lactate or malate it is essential to prevent the reverse reaction by removing the keto acids formed on oxidation.

The lactic enzyme catalyses the oxidation of $l(+)$-lactate to pyruvate. The $d(-)$-enantiomorph[25] is not attacked. No thorough study of the series of α-hydroxy acids has yet been made. α-Hydroxybutyrate is attacked, though much more slowly than lactate. There is indirect[24] evidence that phenyl-substituted hydroxy acids such as p-hydroxyphenyllactate are oxidized in presence of the lactic enzyme. Lactamide[25], malonate, tartronate, glycerate, glycollate, mandellate, gluconate, α-glycerophosphate, citrate, serine and *iso*serine are inactive as substrates. The malic enzyme(26) specifically catalyses the oxidation of $l(-)$-malate to oxaloacetate.

The question of the identity of the lactic and malic enzymes has been under discussion for some time. On the one hand there is evidence that the activity ratio lactic/malic is variable(25, 26), depending upon the source of the enzyme preparation. On the other hand, the similarity of the kinetics are in line with the view that only one enzyme is involved(27). It is possible to obtain from baker's yeast an enzyme preparation which catalyses the oxidation by coenzyme I of malate but not lactate. A malic enzyme of the pyridinoprotein category obtained from *Bact. coli* has also been found to be inactive towards lactate. Recently Straub has purified and crystallized the lactic enzyme of skeletal muscle and shown that it does not attack malate.

The rates of oxidation of lactate or malate are proportional to the concentration of coenzyme within the limiting or saturation concentration (*ca. M/*2000). The half-speed concentration of coenzyme is about $M/20,000$(25, 26). The effect of coenzyme is easily detectable as far down as $M/2,000,000$.

The distribution of the lactic and malic enzymes is very general in animal tissues. Heart is usually the richest source of both enzymes. Brain, kidney, skeletal muscle, blood and liver also contain appreciable concentrations of lactic and malic enzyme.

Fumarate is oxidized quantitatively to oxaloacetate in presence of the malic enzyme. Actually fumarate(26, 28) as such is not oxidized. It is first converted into $l(-)$-malate under the influence of fumarase, a hydrolytic enzyme which is usually found in association with the malic enzyme. The reaction fumarate $\rightleftharpoons l(-)$-malate is reversible, the equilibrium point being reached when the ratio malate/fumarate is about 3.

β-Hydroxybutyric

The enzyme is prepared from the heart muscle of pig(13) by grinding minced and washed muscle with sand and buffer and precipitating the enzyme from the extract with dilute acetic acid. It is much less stable than the lactic or malic enzymes and its distribution is limited to relatively few tissues, notably heart, kidney and liver. The properties and kinetics are similar to those of the lactic or malic systems, with one striking difference. Whereas traces of keto acid formed on oxidation inhibit completely the aerobic oxidation of either lactate or malate, the oxidation of β-hydroxybutyrate is relatively insensitive to the presence of the corresponding keto acid (13). The use of reagents which can fix ketones as fast as formed is obligatory for the aerobic study of the lactic and malic systems, but not for the β-hydroxybutyric system.

The β-hydroxybutyrate enzyme catalyses only the oxidation of the $l(-)$-enantiomorph to acetoacetic acid as well as the converse reduction (13). β-Hydroxypropionic acid is not attacked by the same enzyme. There are no data available whether higher homologues can be oxidized.

Alcohol (yeast)

Negelein & Wulff (29) have succeeded in crystallizing the alcohol enzyme of yeast by the following method. Maceration juice of bottom yeast is heated to 55° to denature inactive protein. The centrifuged solution is made 35 % with respect to acetone. The precipitate is discarded. The enzyme is then brought down with 50 % acetone and dried in vacuo. Successive precipitation of the enzyme first with 62 % and then with 50 % saturated $(NH_4)_2SO_4$ eliminates the more soluble protein impurities. Then follows successive precipitation of the enzyme with 64 %, 60 % and 26 % alcohol at pH 5. At the lowest alcohol concentration the precipitation is very slow and is complete only after 15 hr. at 0°. After the alcohol purification the enzyme can be crystallized by slowly adding saturated $(NH_4)_2SO_4$ to a 1·2 % solution of the enzyme. Crystallization begins when the solution is about 27 % saturated with respect to $(NH_4)_2SO_4$.

The crystalline alcohol enzyme is colourless. It is stable for long periods at 0° when dry. The enzyme is extremely sensitive to traces of heavy metals, particularly copper. The inhibition by heavy metals can be guarded against by addition either of glycine or yeast gum.

One molecule of the enzyme (assuming a molecular weight of

70,000) catalyses in presence of excess ethyl alcohol the reduction of 17,000 molecules of coenzyme I per min. at 20°.

$$\text{alcohol} + \text{coenzyme I} \xrightarrow[\text{T.N.} = 28,500]{\text{T.N.} = 17,000} \text{acetaldehyde}$$
$$+ \text{dihydrocoenzyme I}$$

The turnover number (T.N.) of the oxidation of dihydrocoenzyme I by acetaldehyde (excess) is 28,500. At low concentrations of substrate, aldehyde is reduced (i.e. dihydrocoenzyme I is oxidized) 140 times faster than alcohol is oxidized (i.e. coenzyme I is reduced).

Negelein & Wulff deduce from kinetic considerations that the oxidation of alcohol to aldehyde involves compound formation between the enzyme and coenzyme (see equations on p. 43). The dissociation constant for the compound of enzyme and coenzyme I is 3×10^{-5} M, whereas that for the compound of enzyme and dihydrocoenzyme I is 9×10^{-5} M. In other words, the affinity of coenzyme I for the enzyme is three times that of dihydrocoenzyme I. The enzyme has a much higher affinity for aldehyde than for alcohol (Km aldehyde $= 0.0001$ M, Km alcohol $= 0.024$ M).

A mixture of enzyme and coenzyme I in absence of alcohol reacts very slowly with reduction of the coenzyme. The rate of reduction of the coenzyme by the enzyme is extremely small compared to the rate of reduction by the enzyme in presence of alcohol. Negelein & Wulff attribute this reduction to —SH groups in the enzyme molecule.

The dehydrogenase catalyses the oxidation of simple primary or secondary aliphatic alcohols to the corresponding aldehydes or ketones. Apparently hydroxy acids, sugars and polyhydroxy compounds are not attacked. Physiologically the alcohol enzyme catalyses both the aerobic oxidation of alcohol and the anaerobic reduction of aldehyde.

The reaction between the coenzyme and the alcohol system is in favour of the reduction of aldehyde to alcohol and the oxidation of reduced coenzyme. Hence in order to study the aerobic oxidation of alcohol, it is essential to fix the aldehyde with a suitable ketone-fixing reagent.

Warburg[9] and Euler[10] have demonstrated spectrophotometrically the reduction of coenzyme by alcohol, and the oxidation of reduced coenzyme by aldehyde. In fermenting yeast, aldehyde is reduced to alcohol, whereas some other substance (presumably triosephosphate) is simultaneously oxidized. The coenzyme thereby acts as a link between two pyridinoprotein systems—one acting as the reducing agent for the coenzyme, the other as the oxidizing agent. The net result is the reduction of aldehyde by triosephosphate.

Alcohol (liver)

Lutwak-Mann[30] has made the most comprehensive contribution to our knowledge of the alcohol enzyme of animal tissues. The following is the preparation of Lutwak-Mann. Minced horse liver is mixed with an equal volume of water. After 2 hr. the mixture is centrifuged. The supernatant fluid is heated 15 min. at 52° (precipitate discarded), cooled and mixed with 3 vol. acetone. The precipitate is centrifuged and resuspended in water. The insoluble material is discarded and the extract is mixed with 2 vol. acetone. The dried powder is stable over a period of weeks.

In presence of the enzyme and coenzyme I, ethyl alcohol is oxidized to acetaldehyde. Propyl and amyl alcohol are oxidized as rapidly as ethyl alcohol, whereas methyl alcohol is oxidized comparatively slowly. Glycerol and α-glycerophosphate are inactive as substrates.

Alcohol dehydrogenase of liver is not inhibited by $M/100$ iodoacetic acid, although its counterpart in yeast is inhibited by $M/1000$ iodoacetic acid. The non-identity of the liver and yeast alcohol enzyme is clearly indicated. For the same reason the identity of the aldehyde mutase of liver and the alcohol enzyme is excluded.

Glucose

Harrison[31], in 1931, described in liver an enzyme which catalysed the oxidation of glucose to gluconic acid. He demonstrated conclusively the participation of a coenzyme in this oxidation. Later Adler & Euler[32] showed that Harrison's coenzyme was identical with coenzyme I. The enzyme[31,33] is prepared as follows. Acetone-dried liver is ground up with water. The mixture is filtered, centrifuged and then dialysed. The pH is adjusted to 5·7 and the inactive precipitate discarded. The enzyme is precipitated by addition of $(NH_4)_2SO_4$. Precipitation with $(NH_4)_2SO_4$ is repeated and the precipitate is dried and stored *in vacuo*.

The enzyme specifically catalyses the oxidation of d-glucose (31) to gluconic acid. No other sugar has been found to be active as substrate. The enzyme shows a low affinity for its substrate. Only in presence of massive concentrations of substrate (*ca.* $M/4$) does the enzyme attain maximum activity. The Michaelis constant is 0·07 M.

Das (34) has demonstrated that the glucose dehydrogenase can function in presence of either coenzyme I or coenzyme II. This is one of the two cases recorded of coenzyme unspecificity.

l(+)-*Glutamic* (*animal tissues*)

Adler *et al.* (17, 35) and Dewan (36) have studied in detail the properties of an enzyme found principally in liver and kidney which catalyses the oxidation of *l*(+)-glutamic acid by coenzyme I. The enzyme is conveniently prepared from acetone liver or kidney by acid precipitation of the dialysed aqueous extract and resuspension of the centrifuged precipitate in phosphate buffer.

The enzyme acts specifically on *l*(+)-glutamic acid (i.e. the natural amino acid). It is interesting to note that a different type of enzyme is involved in the oxidation of *d*(−)-glutamic acid (cf. the amino-acid oxidase in Chapter IV). The *l*(+)-enzyme requires the presence of the adeninepyridine dinucleotide, coenzyme I, whereas the *d*(−)-enzyme requires the adenineflavin dinucleotide.

Glutamic acid is oxidized to α-ketoglutaric acid and NH_3. Presumably the first oxidation production is the imino acid which spontaneously hydrolyses into α-ketoglutaric acid and NH_3.

$$COOHCH_2CH_2CHNH_2COOH + \text{coenzyme I} \xrightarrow{\text{enzyme}}$$
$$COOHCH_2CH_2C : NHCOOH + \text{reduced coenzyme I}$$
$$COOHCH_2CH_2C : NHCOOH + H_2O \rightarrow$$
$$COOHCH_2CH_2COCOOH + NH_3$$

Both Adler *et al.*(35,17) and Dewan(36) have shown that this reaction is reversible. By starting with a mixture of reduced coenzyme I, NH_3 and α-ketoglutaric acid, they were able to demonstrate the "synthesis" of glutamic acid. Dewan used the β-hydroxybutyric system for reducing the coenzyme, whereas Adler *et al.* used the alcohol system. The two net reactions are

(1) β-hydroxybutyric acid + NH_3
+ α-ketoglutaric acid $\xrightarrow{\text{coenzyme}}$ acetoacetic acid + glutamic acid

(2) alcohol + NH_3 +

α-ketoglutaric acid $\xrightarrow{\text{coenzyme}}$ aldehyde + glutamic acid

The fact that α-ketoglutaric acid can be reductively aminated to form glutamic acid suggests that the following equilibrium exists:

$$COOHCH_2CH_2C : NHCOOH \rightleftharpoons COOHCH_2CH_2COCOOH + NH_3$$

It is highly probable that this type of reaction accounts for the synthesis of some of the amino acids in the cell.

Krebs & Cohen(65) have carried out experiments with sliced kidney cortex and minced heart muscle which show that the synthesis of glutamic acid from α-ketoglutaric acid and NH_3 is normally linked with the oxidation of α-ketoglutaric acid to succinic acid and CO_2.

α-Ketoglutaric acid + α-iminoglutaric acid + H_2O →
glutamic acid + succinic acid + CO_2

Euler *et al.*(17) find that the glutamic enzyme of liver can collaborate either with coenzyme I or II with equal efficiency. The glutamic enzyme of yeast is specific for coenzyme II, whereas the glutamic enzyme of higher plants is specific for coenzyme I. There is the possibility that the liver enzyme is a mixture of two enzymes each specific for one of the two pyridine nucleotides.

Aldehyde mutase

We owe our first understanding of mutases to the work of Dixon & Lutwak-Mann(21) on the aldehyde mutase (cf. Review of Dixon(22)). It will be recalled that a mutase catalyses the dismutation of the substrate, i.e. an oxido-reduction between two molecules of the substrate, whereby one is oxidized and the other reduced. It has been tacitly assumed in the literature for more than a decade that the so-called aldehyde mutase of liver was identical with the aldehyde enzyme of milk. This was shown to be incorrect. The aldehyde mutase of liver has none of the properties of the aldehyde dehydrogenase of milk. All the available evidence speaks in favour of the aldehyde mutase being a single enzyme.

The enzyme is prepared from horse or pig liver by the following method (21). The minced tissue is mixed with water and allowed to stand for 1 hr. The suspension is centrifuged, the sediment being discarded. The supernatant is mixed with an equal volume of alcohol. The precipitate is discarded. The supernatant is then mixed with 3 vol. of an acetone-ether mixture. The dried precipitate is redissolved in water. The solution is dialysed, centrifuged, and the supernatant fluid heated for 10 min. at 55°. The precipitate is discarded and the supernatant mixed with the acetone-ether mixture as above. The precipitate contains the enzyme.

The dismutation can be followed easily with the aid of the manometric method. The oxidation of aldehyde to acid if carried out in a bicarbonate buffer solution will lead to the liberation of CO_2. The rate of CO_2 evolution is a direct measure of the rate of the oxidation. For each molecule of aldehyde oxidized one molecule of CO_2 is liberated.

The aldehyde mutase catalyses the dismutation of many aliphatic aldehydes (21). Acetaldehyde, propionaldehyde and butyraldehyde are most rapidly attacked. The higher members of the series are less readily dismuted. Aromatic aldehydes such as benzaldehyde and piperonal are attacked extremely slowly if at all. Definite stereochemical specificity is shown in the case of glyceric aldehyde. The dl form is dismuted at one-fifth the rate of the d, suggesting that the l form is both unattacked and inhibitory.

The presence of coenzyme I is necessary for the dismutation and the velocity is proportional to the concentration within the limit of the saturation concentration.

$M/100$ iodoacetate completely inhibits the activity of the aldehyde mutase. This sensitivity to iodoacetate is shown by only a few of the pyridinoproteins and indicates some special grouping, such as a sulphydryl group, in the protein.

The preparation of aldehyde mutase catalyses not only the dismutation of aldehydes but also the oxidation of aldehydes by a whole series of α-ketonic acids(24). Pyruvate, oxaloacetate, mesoxalate, α-ketoglutarate, α-ketobutyrate and α-keto*iso*caproate can act as oxidants. In the presence of pyruvate or oxaloacetate, the aldehyde oxidized to acid is equivalent to the ketonic acid reduced. In other words, the dismutation is completely suppressed when suitable α-ketonic acids are present. Aldehyde as oxidant is quantitatively replaced by ketonic acid. Not all α-ketonic acids are as efficient as pyruvate or oxaloacetate. The less active ketonic acids merely compete with aldehyde as oxidant but do not suppress altogether the dismutation. Acetoacetic acid (β-ketonic) cannot act as oxidant. Presumably the lactic and malic as well as related enzymes present in the preparation of aldehyde mutase are necessary for catalysing the reduction of the various ketonic acids by reduced coenzyme I.

No thorough study has yet been made of the distribution of the aldehyde mutase. Liver is the richest known source of the enzyme and kidney less so. There is apparently none in skeletal or heart muscle. There is evidence that the enzyme is present in brain.

Glycolysis and fermentation

The following are at present believed to be the key reactions in the breakdown of hexose to lactic acid in animal tissues and of hexose to alcohol and CO_2 in yeast:

(1) hexosediphosphate $\overset{\text{zymohexase}}{\rightleftharpoons}$ dihydroxyacetonephosphate
$$+ \text{3-phosphoglyceraldehyde}$$

(2) dihydroxyacetonephosphate $\overset{\text{isomerase}}{\rightleftharpoons}$ 3-phosphoglyceraldehyde

(3) 3-phosphoglyceraldehyde + phosphate
$$+ \text{coenzyme I} \overset{\text{triosephosphoric enzyme}}{\rightleftharpoons} \text{1 : 3-diphosphoglycerate}$$
$$+ \text{dihydrocoenzyme I}$$

(4) 1 : 3-diphosphoglycerate + adenosinediphosphate
$$\rightleftharpoons 3\text{-phosphoglycerate} + \text{adenosinetriphosphate}$$

(5) 3-phosphoglycerate $\xrightleftharpoons{\text{phosphoglyceromutase} + \text{enolase}}$ phosphopyruvate

(6) phosphopyruvate + adenosinediphosphate
$$\rightarrow \text{pyruvate} + \text{adenosinetriphosphate}$$

Muscle glycolysis

(7) dihydrocoenzyme I + pyruvate
$$\xrightleftharpoons{\text{lactic enzyme}} \text{coenzyme I} + \text{lactate}$$

Yeast fermentation

(8) pyruvate $\xrightarrow{\text{carboxylase}}$ acetaldehyde + CO_2

(9) dihydrocoenzyme I + acetaldehyde
$$\xrightleftharpoons{\text{alcohol enzyme}} \text{coenzyme I} + \text{alcohol}$$

Thus the fundamental oxido-reductions of glycolysis and fermentation are:

Muscle glycolysis

(10) 3-phosphoglyceraldehyde + pyruvate + phosphate
$$\rightleftharpoons 2 : 3\text{-phosphoglycerate} + \text{lactate}$$

Yeast fermentation

(11) 3-phosphoglyceraldehyde + aldehyde + phosphate
$$\rightleftharpoons 2 : 3\text{-phosphoglycerate} + \text{alcohol}$$

In addition to the above-mentioned oxido-reductions the following reactions have been demonstrated in glycolysing extracts of skeletal muscle (23, 27):

$$\text{α-glycero-phosphate} + \begin{cases} \text{pyruvate} \\ \text{oxaloacetate} \end{cases} \xrightarrow[\text{coenzyme I}]{\text{enzymes}} \text{triosephosphate} + \begin{cases} \text{lactate} \\ \text{malate} \end{cases}$$

$$\left.\begin{array}{l}\text{glyceraldehyde} \\ \text{glycollic aldehyde}\end{array}\right\} + \begin{cases} \text{pyruvate} \\ \text{oxaloacetate} \end{cases} \xrightarrow[\text{coenzyme I}]{\text{enzymes}} \left.\begin{array}{l}\text{glyceric acid} \\ \text{glycollic acid}\end{array}\right\} + \begin{cases} \text{lactate} \\ \text{malate} \end{cases}$$

The α-glycerophosphate and triose enzymes catalyse the reduction of coenzyme I by their respective substrates, and the lactic and malic enzymes catalyse the oxidation of reduced coenzyme I by their respective substrates.

These oxido-reductions by no means exhaust the possibilities of interaction in aqueous extracts of minced skeletal muscle. It is important to note that under certain conditions many of these reactions can proceed in the reverse direction.

The preparation of the glycolysing extract of skeletal muscle is made as follows (24). The skeletal muscles of a rabbit are cooled to 0° and minced. The mince is mixed with 2 vol. of iced water. After 30 min. the insoluble material is filtered through muslin and the filtrate is mixed with excess acetone. The dried precipitate is resuspended and the mixture is dialysed. The centrifuged solution contains active triosephosphoric, triose, α-glycerophosphoric, lactic and malic enzymes as well as zymohexase, isomerase, and enolase. Coenzyme I has to be added to the extract.

Zymohexase

Meyerhof & Lohmann (66, 69) discovered in skeletal muscle of rabbit an enzyme which catalyses the reversible' splitting of fructose-1 : 6-diphosphate into a molecule each of glyceraldehyde-phosphate and dihydroxyacetonephosphate:

$$
\begin{array}{ccc}
\mathrm{H_2COPO_3H_2} & & \\
| & & \\
\mathrm{CO} & \overset{\mathrm{O}}{\underset{}{\mathrm{C}}}\diagdown\mathrm{H} & \mathrm{CH_2OH} \\
| & | & | \\
\mathrm{HOCH} \rightleftharpoons & \mathrm{HCOH} \quad + & \mathrm{C\!=\!O} \\
| & | & | \\
\mathrm{HCOH} & \mathrm{H_2COPO_3H_2} & \mathrm{H_2COPO_3H_2} \\
| & & \\
\mathrm{HCOH} & & \\
| & & \\
\mathrm{H_2COPO_3H_2} & & \\
\end{array}
$$

| fructose-1: 6-diphosphoric acid | 3-phosphoglyceraldehyde | dihydroxyacetonephosphoric acid |

The reaction is reversible, but the measurement of the equilibrium constant is complicated by the fact that crude preparations of zymohexase contain an enzyme which catalyses the reversible transformation of glyceraldehydephosphate into dihydroxyacetonephosphate. By addition of ketone fixatives such as cyanide, hydrazine, etc. the scission or forward reaction can be made to go to completion.

Meyerhof *et al.*(67) found that the same enzyme catalyses the aldol condensation of dihydroxyacetonephosphate (dap) with various aldehydes, e.g.

dap + acetaldehyde → methyltetrosephosphate
dap + *d*-glyceraldehyde → fructose-1-phosphate

Other condensations were obtained with formaldehyde, glycollic aldehyde, methylglyoxal, etc. Glyceraldehydephosphate cannot replace "dap" in these reactions.

Zymohexase is widely distributed in animal and plant tissues and in micro-organisms. It may well be considered as part of the fundamental enzymic equipment of practically all cells.

Herbert *et al.*(68) have isolated zymohexase of rabbit skeletal muscle in homogeneous form and studied some of its properties and kinetics. It is present in high concentration in rabbit skeletal muscle and accounts for some 5% of the total water soluble protein. Assuming a molecular weight of 100,000 each molecule of zymohexase would catalyse the transformation of 210 molecules of hexosediphosphate per min. at 38°.

Isomerases

Meyerhof & Lohmann(69) described an enzyme in rabbit skeletal muscle which catalyses the transformation:

dihydroxyacetonephosphate ⇌ 3-phosphoglyceraldehyde

Under physiological conditions the equilibrium is greatly in favour of "dap" formation (97 %). The same enzyme probably catalyses the reversible conversion of 6-phosphoglucose into 6-phosphofructose.

Meyerhof & Kiessling(70), in experiments on muscle extracts, have obtained evidence for the existence of two other isomerases which catalyse the following successive rearrangements of 3-phosphoglycerate:

$$
\begin{array}{ccccc}
\text{COOH} & & \text{COOH} & & \text{COOH} \\
| & & | & & | \\
\text{HCOH} & \rightleftharpoons & \text{HCOPO}_3\text{H}_2 & \rightleftharpoons & \text{COPO}_3\text{H}_2 \\
| & & | & & \| \\
\text{H}_2\text{COPO}_3\text{H}_2 & & \text{H}_2\text{COH} & & \text{H}_2\text{C}
\end{array}
$$

3-phosphoglyceric   2-phosphoglyceric   phosphopyruvic

Triosephosphoric enzyme

Our present knowledge of the detailed reactions of muscle glycolysis and yeast fermentation arises mainly from the contributions of the Meyerhof and Embden schools. The study of the oxidation enzymes involved in these processes was later undertaken by Euler, Warburg and others. The pivotal position of triosephosphate was at once appreciated by the various workers in the field. At first it was generally assumed that the oxidation of 3-phosphoglyceraldehyde to 3-phosphoglycerate by coenzyme I was a direct reaction. Warburg & Christian (64, 71), however, showed that this reaction took place in two stages:

$$
\begin{array}{ll}
\text{H} & \text{OPO}_3\text{H}_2 \\
\text{C}=\text{O} & \text{C}=\text{O} \\
| & | \\
\text{HCOH} + \text{H}_3\text{PO}_4 + \text{coenzyme I} \rightleftharpoons \text{HCOH} + \text{dihydrocoenzyme I} \\
| & | \\
\text{HCOPO}_3\text{H}_2 & \text{HCOPO}_3\text{H}_2 \\
\text{H} & \text{H}
\end{array}
$$

3-phosphoglyceraldehyde   1 : 3-diphosphoglyceric

$$
\begin{array}{ll}
\text{OPO}_3\text{H}_2 & \text{OH} \\
\text{C}=\text{O} & \text{C}=\text{O} \\
| & | \\
\text{HCOH} + \text{adenosinediphosphate} \rightleftharpoons \text{HCOH} + \text{adenosinetriphosphate} \\
| & | \\
\text{HCOPO}_3\text{H}_2 & \text{HCOPO}_3\text{H}_2 \\
\text{H} & \text{H}
\end{array}
$$

1 : 3-diphosphoglyceric   3-phosphoglyceric

The triosephosphoric enzyme catalyses only the initial oxidation. The subsequent formation of 3-phosphoglycerate involves a different enzyme.

Warburg & Christian (71) have isolated the triosephosphoric enzyme of baker's yeast in crystalline form by the following method. Maceration juice of dried baker's yeast is made 55 % with respect to acetone at 10°. The precipitate is rubbed up with water and the suspension is heated to 55° for 10 min. The precipitate of denatured protein is discarded. The enzyme is then precipitated at pH 4·5 as the salt of nucleic acid. The precipitate is washed twice with acetate buffer of pH 4·7 and then dissolved in water by cautious addition of dilute soda (final pH 6). Treatment with Al(OH)₃ to remove coloured impurities is finally followed by

15 min. heating at 60°. The enzyme at this stage can be obtained crystalline by addition of NH_3 to a half-saturated $(NH_4)_2SO_4$ solution of the protein.

The crystallized enzyme is stable at 0° in solution but is partially destroyed (40 %) in the course of drying. It is extremely sensitive to the presence of heavy metals.

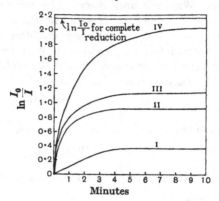

Fig. 9. The effect of the concentration of phosphate on the rate of reduction of coenzyme I by the triosephosphoric enzyme system.

I 0.87×10^{-6} M phosphate per ml., pH 7·4.

II 1.68×10^{-5} M phosphate per ml., pH 7·4.

III 3.36×10^{-5} M phosphate per ml., pH 7·4.

IV 3.36×10^{-5} M phosphate per ml., pH 8·45.

I_0/I is the light diminution at 340 mμ.; temp. 20°; $d = 0.557$ cm. (Warburg & Christian [1939]. *Biochem. Z.* 303, 64.)

A mixture of 3-phosphoglyceraldehyde, the enzyme and co-enzyme I does not react to any appreciable degree. On addition of inorganic phosphate reduction of the coenzyme rapidly takes place (cf. Fig. 9). The velocity and extent of reduction of the coenzyme is proportional to the concentration of phosphate. Thus two facts emerge: (1) that phosphate is an essential reactant and (2) that the reaction is reversible and that the equilibrium lies to the left in presence of small amounts of phosphate (*ca.* 0.87×10^{-6} M) and to the right in presence of comparatively larger amounts (*ca.* 3.36×10^{-5} M).

Warburg & Christian(71) assume that phosphate combines with 3-phosphoglyceraldehyde to form 1 : 3-diphosphoglyceraldehyde. Whether this combination takes place spontaneously or is catalysed by the enzyme is still in doubt. There is in fact no evidence that 1 : 3-diphosphoglyceraldehyde is even formed as an intermediary. The product of oxidation has been isolated by Negelein & Brömel(63,72) and shown to be a diphospho derivative of glyceric acid with the probable constitution 1 : 3-diphosphoglyceric acid.

Dihydrocoenzyme I is oxidized by 1 : 3-diphosphoglycerate in presence of the enzyme practically to completion. The equilibrium is thus clearly in favour of the oxidation of the dihydrocoenzyme rather than the reduction of the coenzyme. Neither 3-phosphoglycerate nor phosphopyruvate can replace 1 : 3-diphosphoglycerate as oxidant.

Assuming that the molecular weight of the enzyme is 100,000 calculation shows that one molecule catalyses the reduction of 16,600 molecules of coenzyme I per min. at pH 7·4 and 20°; at pH 8·45 the turnover number is 28,500.

Arsenate can replace phosphate in the catalytic system. Presumably arseno-3-phosphoglyceraldehyde is first formed and this becomes oxidized to the corresponding 1-arseno-3-phosphoglycerate. The product of oxidation is unstable and decomposes into arsenate and 3-phosphoglycerate. The oxidation of 3-phosphoglyceraldehyde to 3-phosphoglycerate goes to completion in presence of arsenate, since the instability of the oxidation product prevents the attainment of an equilibrium state. For kinetic measurements the use of arsenate instead of phosphate has obvious advantages.

The oxido-reductions between 3-phosphoglyceraldehyde on the one hand and pyruvate and oxaloacetate on the other have been studied by Green et al.(24), using dialysed extracts of the "acetone powder" of rabbit skeletal muscle as the source of the various enzymes required. The mechanism of these oxido-reductions was shown to involve a cycle of reduction of coenzyme I by triosephosphate and oxidation of dihydrocoenzyme I by either of the two α-ketonic acids. In these experiments no inorganic phosphate was added. Yet a mixture of triosephosphate and pyruvate reacted to completion. That is to say, the equilibrium was entirely in favour of the production of lactate and 8-phosphoglycerate. Green *et al.*(24) were able to reverse the reaction by addition of cyanide to a mixture of lactate, phosphoglycerate, coenzyme I and the glycolysing extract. Cyanide reacts almost instantaneously with the pyruvate and triosephosphate, which are formed in small quantity, and thereby eliminates the forward reaction. Here again addition of inorganic phosphate was not necessary. This can mean either that phosphate was not necessary in these reactions or more likely that traces of phosphate were undergoing cyclical phosphorylation and dephosphorylation. The nature of this cycle has yet to be elucidated.

The triosephosphoric enzyme is found principally in brain, skeletal and cardiac muscle. There is some activity in intestine, though practically none in liver and kidney. The enzyme is also widely distributed in yeast and bacteria.

Rapkine(48) has treated the triosephosphoric enzyme with reagents known either to reduce or oxidize —SS— and —SH groups respectively. The results of these experiments have led him to the conclusion that —SH groups are intimately connected with the activity of the enzyme and that mild interference with these groups leads to loss of activity.

Triose enzyme

The glycolysing solution prepared as described above from rabbit muscle contains an enzyme(24) which catalyses the oxidation of glyceraldehyde ($CH_2OHCHOHCHO$) or glycollic aldehyde (CH_2OHCHO) by coenzyme I.

The optimum pH for the triose enzyme is about 8·0. The Km for dl-glyceraldehyde is $M/250$. The Km of the coenzyme is $M/70,000$. Glyceraldehyde is attacked somewhat more rapidly than glycollic aldehyde. Tetrose (erythrose) is hardly oxidized, while pentoses, hexoses, etc. are completely inactive as substrates. Dihydroxyacetone cannot replace glyceraldehyde. Only one optical isomer of glyceraldehyde is oxidized, viz. the $l(+)$-isomer. The unnatural isomer does not inhibit the oxidation of the natural isomer.

The triose enzyme(24) of skeletal muscle is not identical with the aldehyde mutase of liver. The former acts specifically on triose and diose, whereas the latter acts unspecifically on a whole series of aliphatic aldehydes. Furthermore, the dismutation of glyceraldehyde is much more rapid with the aldehyde mutase than is the oxidation with the triose enzyme.

The fact that two different enzymes catalyse the same reaction makes it difficult to ascertain the respective distributions of the two enzymes concerned. The available data show that the triose enzyme is present principally in skeletal and cardiac muscle.

The triose enzyme, like the triosephosphoric, is extremely sensitive to iodoacetic acid (24). Even $M/3000$ concentration is sufficient to inhibit the activity of the enzyme completely. There is a slight lag period of some 2 min. before the inhibitory action of iodoacetic acid begins. The length of the lag period is increased if the enzyme is incubated with the substrate prior to the introduction of iodoacetic acid.

α-Glycerophosphoric enzyme

The glycolysing extract from rabbit skeletal muscle contains an enzyme which catalyses the oxidation of α-glycerophosphate by coenzyme I. The work of Euler and his colleagues (38) has shown that this enzyme is a typical pyridinoprotein. It differs from the triose and triosephosphoric enzymes in its insensitivity to iodoacetic acid.

When α-glycerophosphate is oxidized by α-ketonic acids in extracts of skeletal muscle, phosphoglycerate is isolated as the end product (24). The evidence is in favour of the view that the reaction occurs in two stages:

α-glycerophosphate + pyruvate → triosephosphate + lactate
triosephosphate + pyruvate → phosphoglycerate + lactate

The α-glycerophosphoric enzyme catalyses only the first reaction, whereas the triosephosphoric enzyme, which is always present concomitantly, catalyses the further oxidation of triosephosphate to phosphoglycerate.

The $d(-)$-isomer only of α-glycerophosphate is attacked. The $l(+)$-isomer and β-glycerophosphate have no activity as substrates.

The distribution of the enzyme in animal tissues is restricted to a few tissues, notably skeletal muscle, intestine and heart muscle.

Formic

Adler & Sreenivasaya[39] have examined systematically the properties of a formic enzyme which they prepared from dried peas. The steps in the preparation of the enzyme include (1) grinding and extracting with phosphate, (2) precipitating the filtrate with 60 % $(NH_4)_2SO_4$, (3) redissolving the dried precipitate in water, dialysing, and centrifuging off inactive material. The enzyme catalyses the oxidation of formate to CO_2 and H_2O. Coenzyme I is essential for the reaction. The reaction between formate and coenzyme I proceeds practically to completion. That is to say, the equilibrium is entirely in favour of the reduction of the coenzyme and oxidation of formate. Adler[39] failed to reverse this reaction by bubbling CO_2 into a solution containing reduced coenzyme and the enzyme. There is the possibility that CO_2 is not the first product of oxidation and that the conversion of the hypothetical intermediary to CO_2 and H_2O is an irreversible stage.

TRIPHOSPHOPYRIDINOPROTEIN ENZYMES

There are at least four known representatives of this class, the hexosemonophosphoric, phosphogluconic, isocitric and glutamic enzymes. They all catalyse the oxidation of their various substrates by coenzyme II. As yet these reactions, with the exception of the oxidation of glutamate, have not been reversed, i.e. reduced coenzyme II does not react with the oxidized substrates. Comparatively little is known of the role of coenzyme II systems in cellular respiration.

Hexosemonophosphoric enzyme

Warburg & Christian[40, 41] have prepared active enzymes from both red blood corpuscles and yeast which catalyse the oxidation of glucose-6-phosphate

$$CH_2OPO_3H_2(CHOH)_4CHO$$

to 6-phosphogluconate

$$CH_2OPO_3H_2(CHOH)_4COOH.$$

Recently Negelein(42) has succeeded in purifying the yeast enzyme to an extremely high degree.

Lebedev juice of bottom yeast is acidified to pH 4·6 and kept at 0° for 15 hr. The precipitate is discarded. The solution is brought to pH 9 and warmed to 30°. After 6 hr. the solution is cooled and centrifuged. The pH is then adjusted to 4·8 and the solution is ¼th saturated with $(NH_4)_2SO_4$. The precipitate is discarded. With 70 % saturation of $(NH_4)_2SO_4$ the enzyme is precipitated. The precipitate is redissolved in water and made 30 % saturated with respect to $(NH_4)_2SO_4$. After warming to 40° for 50 min. the inactive precipitate is discarded. The solution is then dialysed, acidified to 4·8 at 0° and made 10 % with respect to ethyl alcohol. The enzyme is precipitated out under these conditions. The precipitate is dissolved in very dilute bicarbonate solution. By reacidifying to pH 4·8, the enzyme is precipitated without addition of either alcohol or $(NH_4)_2SO_4$. The isoelectric point of the enzyme is 4·8. But only after the bulk of impurities is removed does the enzyme become insoluble at the isoelectric point. The final precipitate is dried and kept *in vacuo*. The yield is 8·5 %, calculated on the basis of the enzyme content in the original Lebedev juice.

The purified enzyme contains 13·5 % N(42). It is soluble in water except at the isoelectric hydrogen-ion concentration. Solutions of the purified enzyme are very unstable—destruction is complete in 24 hr. at 0°. Stability is of course much greater in the unpurified state. Yeast polysaccharide exerts a protective action. The purified enzyme is most stable in 20–30 % $(NH_4)_2SO_4$.

The enzymic oxidation of hexosemonophosphate depends uniquely upon the presence of coenzyme II. Coenzyme I is completely inactive. As with all coenzyme systems, hydrogen is transferred from hexosemonophosphate to the coenzyme. The reversal of the oxidation, i.e. the reduction of phosphohexonate to hexosemonophosphate by reduced coenzyme, has not been demonstrated.

The distribution of the hexosemonophosphoric enzyme in animal tissues is not known. The red blood corpuscles of the rat offer a convenient source of supply(70). The corpuscles are thrice washed with 0·9 % NaCl, cytolysed with a small amount of distilled water and the solution is kept at 0°. The crystalline precipitate of haemoglobin carries down with it the enzyme. The precipitate is centrifuged and resuspended in water. The enzyme is thereby eluted.

The following calculations (2) are of interest in connection with the activity of the coenzyme in the hexosemonophosphoric system. In presence of excess enzyme and flavoprotein, and in an atmosphere of pure oxygen, 0·001 mg. of coenzyme II can transfer 1 μl. of oxygen per min. The Q_{O_2} of the coenzyme is therefore 60,000. Under optimum conditions 1 mg. of the "pure" enzyme catalyses the uptake of 350 μl. O_2 per min. ($Q_{O_2} = 60,000$) at 38°.

Fructosemonophosphate and hexosediphosphate are not attacked by the enzyme. Phosphogluconate is oxidized in presence of the crude enzyme preparation but not in presence of the highly purified hexosemonophosphoric enzyme, indicating that another enzyme is responsible for its activation.

Phosphogluconic enzyme

Dickens (43), Lipmann (44) and Warburg (45) have recently called attention to an enzyme from yeast which catalyses the oxidation of 6-phosphogluconate to the corresponding ketonic acid. Coenzyme II is involved in this oxidation. Dickens has also observed the enzyme in animal tissues. The first product of oxidation is probably phosphoketohexonate

$$CH_2OPO_3H_2(CHOH)_3COCOOH,$$

which can be further stepwise degraded by other enzymes in the Dickens preparation until a C_4 monocarboxylic dihydroxy acid is formed probably identical with phosphoerythronic acid

$$CH_2OPO_3H_2CHOHCHOHCOOH.$$

The other oxidation enzymes involved also seem to require coenzyme II.

Glutamic

Euler et al. (46, 49) have described an enzyme isolated from bottom yeast which specifically catalyses the oxidation of $l(+)$-glutamic acid by coenzyme II. Oxidative deamination takes place with production of NH_3 and α-ketoglutaric acid. Coenzyme I cannot replace coenzyme II as the specific oxidizing agent for the substrate. It is interesting to note that the corresponding glutamic enzyme of higher plants specifically requires coenzyme I, whereas the liver enzyme does not distinguish between the two pyridine nucleotides.

Isocitric

Martius & Knoop[50] demonstrated in liver mince the formation of α-ketoglutarate from citrate. They postulated the following sequence of events:

$$
\begin{array}{ccc}
\overset{\text{OH}}{\underset{\substack{\text{H}\overset{|}{\text{C}}\text{H} \\ |\\ \text{COOH}}}{\text{COOH C CH}_2\text{COOH}}}
& \xrightarrow{-\text{H}_2\text{O}} &
\underset{\substack{\text{CH} \\ |\\ \text{COOH}}}{\text{COOH C CH}_2\text{COOH}}
& \xrightarrow{+\text{H}_2\text{O}} &
\overset{\text{H}}{\underset{\substack{\text{H}\overset{|}{\text{C}}\text{OH} \\ |\\ \text{COOH}}}{\text{COOH C CH}_2\text{COOH}}} \\
\text{citric} & & \textit{cis}\text{aconitic} & & \text{isocitric}
\end{array}
$$

$$
\xrightarrow{-2\text{H}}
\overset{\text{H}}{\underset{\substack{\text{CO} \\ |\\ \text{COOH}}}{\text{COOH C CH}_2\text{COOH}}}
\xrightarrow{-\text{CO}_2}
\underset{\substack{\text{CH}_2 \\ |\\ \text{CH}_2 \\ |\\ \text{CO} \\ |\\ \text{COOH}}}{\text{COOH}}
$$

$$\alpha\text{-keto-}\beta\text{-carboxyglutaric acid} \qquad\qquad \alpha\text{-ketoglutaric}$$

Martius[61] and Johnson[62] have studied the properties of aconitase, the enzyme (not identical with fumarase?) which catalyses the transformation of citrate to isocitrate. The equilibrium values of the concentrations of the three reactants at neutral pH are as follows:

$$\text{citrate} \xrightarrow{-\text{H}_2\text{O}} \textit{cis}\text{aconitate} \xrightarrow{+\text{H}_2\text{O}} \text{isocitrate}$$
$$\phantom{\text{citrate}}\;80\,\% \qquad\qquad 4\,\% \qquad\qquad 16\,\%$$

The enzyme is widely distributed in animal tissues.

Adler et al.[51,52] have made a systematic investigation of the properties of the enzyme which catalyses the oxidation of isocitrate (but not citrate) to α-ketoglutarate. Acetone-dried heart muscle serves as the starting material of their method of preparation. The aqueous extract of the acetone powder is precipitated at $0°$ with 75 % acetone. After redissolving the precipitate in water the acetone treatment is repeated. The precipitate is dried with acetone and ether. A solution of this stable acetone powder contains an active isocitric enzyme but no aconitase.

The isocitric enzyme catalyses the oxidation of isocitrate by coenzyme II (52). "It is difficult to say whether coenzyme I has an action in this system which is 150 times weaker than that of coenzyme II or if coenzyme I is actually inactive. At any rate the great difference in the degree of action indicates that the isocitric enzyme is practically specific for coenzyme II."

The optimal rate of oxidation takes place when the concentration of the substrate is as low as $1.25 \times 10^{-5}\ M$. Thus the Michaelis constant Km is lower than $1.25 \times 10^{-5}\ M$, one of the lowest recorded values for oxidizing enzymes. The optimum pH for the reaction is $ca.$ 7–7.5. Below pH 6.5 the rate falls off rapidly. Phosphate buffers cannot be used because of the inhibition of the enzyme by the phosphate ion.

The product of oxidation is probably α-keto-β-carboxyglutarate. This substance is unstable and spontaneously decomposes with liberation of CO_2 to form α-ketoglutarate. The latter has been isolated from the reaction mixture as the 2 : 4-dinitrophenylhydrazone.

The reaction isocitrate $\rightarrow \alpha$-ketoglutarate is irreversible (52). There is no means of deciding whether the primary reaction, isocitrate $\rightarrow \alpha$-keto-β-carboxyglutarate, is also irreversible, since the secondary decomposition of the ketonic acid is extremely rapid.

Adler $et\ al.$ (52) observed that at low dilutions the activity of the enzyme is no longer proportional to its concentration. This discrepancy they tracked down to some activator present in minimal concentration in the enzyme preparation. On dilution of the enzyme the concentration of the activator becomes too low for maximum enzymic activity. Analysis showed that Mn and Mg salts had the properties of the activator present in the preparation and their addition to the enzyme solution eliminated the dilution effect. The optimum concentration for Mn^{++} was $5 \times 10^{-4}\ M$, and for Mg^{++} $2.5 \times 10^{-3}\ M$. The maximal activity brought about by Mg^{++} is only 59 % of that caused by Mn^{++}. Ca^{++}, Zn^{++} and Cd^{++} were inactive. The mechanism of activation by Mn^{++} and Mg^{++} is obscure.

Iodoacetate inhibits 97 % and 75 % in concentrations of $M/100$ and $M/1000$ respectively. No inhibition is obtained with cyanide, malonate, fluoride and oxalate.

Heart, liver, kidney and adrenal gland are the richest sources of the isocitric enzyme. Skeletal muscle, lung, brain and testis are relatively weak in the enzyme. Ovary and intestine are intermediate between these two groups.

The end-product of the oxidation of isocitrate, viz. α-ketoglutarate, is also the end-product of the oxidation of glutamate. Since both the oxidation enzymes concerned collaborate with coenzyme II it should be possible to synthesize glutamic acid from isocitrate and NH_3 in presence of the two enzymes and coenzyme II. Adler et al. (52) have shown that this synthesis can in fact take place. Isocitrate reduces coenzyme II to the dihydro form; in turn ammonium α-ketoglutarate (actually the iminoglutarate) is reduced by dihydrocoenzyme II to glutamic acid:

isocitrate + Co II → α-ketoglutarate + CO_2 + reduced Co II

α-ketoglutarate + NH_3 → α-iminoglutarate

α-iminoglutarate + reduced Co II → glutamate + Co II

REFERENCES

(1) Harden & Young [1906]. *Proc. Roy. Soc.* B, 125, 171.
(2) Warburg & Christian [1933]. *Biochem. Z.* 266, 377.
(3) Warburg, Christian & Griese [1935]. *Biochem. Z.* 282, 157.
(4) Euler & Schlenk [1937]. *Hoppe-Seyl. Z.* 246, 64.
(5) Karrer & Warburg [1936]. *Biochem. Z.* 285, 297.
(6) Karrer, Schwarzenbach, Benz & Solmssen [1936]. *Helv. Chim. Acta,* 19, 811.
(7) Euler, Albers & Schlenk [1935]. *Hoppe-Seyl. Z.* 237, 1.
(8) Schlenk, Hellström & Euler [1938]. *Ber. chem. Ges.* 71, 1471.
(9) Warburg & Christian [1936]. *Helv. chim. Acta,* 19, E 79.
(10) Euler, Adler & Hellström [1936]. *Hoppe-Seyl. Z.* 241, 239.
(11) Warburg [1937]. *Ergebn. Enzymforsch.* 7, 210.
(12) Negelein & Wulff [1937]. *Biochem. Z.* 290, 445.
(13) Green, Dewan & Leloir [1937]. *Biochem. J.* 31, 934.
(14) Dewan & Green [1937]. *Biochem. J.* 31, 1074.
(15) Euler, Adler, Günther & Hellström [1937]. *Hoppe-Seyl. Z.* 245, 232.
(16) Euler, Adler & Günther [1937]. *Hoppe-Seyl. Z.* 247, 65.

72   PYRIDINOPROTEIN ENZYMES

(17) Euler, Adler, Günther & Das [1938]. *Hoppe-Seyl. Z.* **254**, 61.
(18) Adler & Günther [1938]. *Hoppe-Seyl. Z.* **253**, 143.
(19) Clark [1928]. *The determination of hydrogen ion concentration*, 3rd edition, Baltimore.
(20) Michaelis [1933]. *Oxydations-Reduktions Potentiale*, 2 Auflage, Berlin.
(21) Dixon & Lutwak-Mann [1937]. *Biochem. J.* **31**, 1347.
(22) Dixon [1939]. *Ergebn. Enzymforsch.* **8**, 217.
(23) Adler & Hughes [1938]. *Hoppe-Seyl. Z.* **253**, 71.
(24) Green, Needham & Dewan [1937]. *Biochem. J.* **31**, 2327.
(25) Green & Brosteaux [1936]. *Biochem. J.* **80**, 1489.
(26) Green [1936]. *Biochem. J.* **30**, 2095.
(27) Das [1937]. *Biochem. J.* **31**, 1116; **31**, 1124.
(28) Laki [1937]. *Biochem. J.* **31**, 1113.
Laki, Straub & Szent-Györgyi [1937]. *Hoppe-Seyl. Z.* **247**, 1.
(29) Negelein & Wulff [1937]. *Biochem. Z.* **289**, 436; **293**, 351.
(30) Lutwak-Mann [1938]. *Biochem. J.* **32**, 1364.
(31) Harrison [1931]. *Biochem. J.* **25**, 1016.
(32) Adler & Euler [1935]. *Hoppe-Seyl. Z.* **232**, 6.
(33) Harrison [1932]. *Biochem. J.* **26**, 1295.
(34) Das [1936]. *Hoppe-Seyl. Z.* **238**, 269.
(35) Adler, Das & Euler [1937]. *Sr. Vet. Akad. Ark. Kemi*, **12**, 1.
(36) Dewan [1938]. *Biochem. J.* **32**, 1378.
(37) Dixon & Zerfas [1940]. *Biochem. J.* **34**, 371.
(38) Euler, Adler & Günther [1937]. *Hoppe-Seyl. Z.* **249**, 1.
(39) Adler & Sreenivasaya [1937]. *Hoppe-Seyl. Z.* **249**, 24.
(40) Warburg & Christian [1931]. *Biochem. Z.* **242**, 206.
(41) Warburg & Christian [1932]. *Biochem. Z.* **254**, 438.
(42) Negelein & Gerischer [1936]. *Biochem. Z.* **284**, 289.
(43) Dickens [1938]. *Biochem. J.* **32**, 1626.
(44) Lipmann [1936]. *Nature, Lond.*, **138**, 588.
(45) Warburg & Christian [1936]. *Biochem. Z.* **287**, 440; [1937]. *Ibid.* **292**, 287.
(46) Euler, Adler & Stenhoff Eriksen [1937]. *Hoppe-Seyl. Z.* **248**, 227.
(47) Quastel & Wheatley [1938]. *Biochem. J.* **32**, 936.
(48) Rapkine [1938]. *Biochem. J.* **32**, 1729.
(49) Adler, Günther & Everett [1938]. *Hoppe-Seyl. Z.* **255**, 27.
(50) Martius & Knoop [1937]. *Hoppe-Seyl. Z.* **246**, 1.
(51) Euler, Adler & Plass [1938]. *Ark. Kemi Min. Geol.* **133**, 1.
(52) Adler, Euler, Günther & Plass [1939]. *Biochem. J.* **33**, 1028.
(53) Laki [1937]. *Hoppe-Seyl. Z.* **249**, 63.
(54) Barron [1939]. *Physiol. Rev.* **19**, 184.
(55) Barron & Hastings [1934]. *J. biol. Chem.* **107**, 567.
(56) Wurmser & Mayer-Reich [1933]. *J. chim. Phys.* **30**, 249.
(57) Euler, Adler, Günther & Hellström [1937]. *Hoppe-Seyl. Z.* **245**, 217.
(58) Adler & Sreenivasaya [1937]. *Hoppe-Seyl. Z.* **294**, 24.
(59) Hoff-Jørgensen [1938]. *Skand. Arch. Physiol.* **79**, 1.
(60) Clark [1938]. *J. Appl. Phys.* **9**, 97.
(61) Martius [1938]. *Hoppe-Seyl. Z.* **242**, 1.
(62) Johnson [1939]. *Biochem. J.* **33**, 1046.
(63) Negelein & Brömel [1939]. *Biochem. Z.* **301**, 135.
(64) Warburg & Christian [1939]. *Biochem. Z.* **301**, 221.
(65) Krebs & Cohen [1939]. *Biochem. J.* **33**, 1895.

(66) Meyerhof & Lohmann [1934]. *Biochem. Z.* **275**, 430.
   Meyerhof [1938]. *Bull. Soc. Chim. biol.* **20**, 1033.
(67) Meyerhof, Lohmann & Schuster [1936]. *Biochem. Z.* **286**, 301; **286**, 319.
(68) Herbert, Gordon, Subrahmanyan & Green [1940]. *Biochem. J.* **34**, 1108.
(69) Meyerhof & Lohmann [1934]. *Biochem. Z.* **271**, 102.
(70) Meyerhof & Kiessling [1935]. *Biochem. Z.* **280**, 99.
(71) Warburg & Christian [1939]. *Biochem. Z.* **303**, 40.
(72) Negelein & Brömel [1939]. *Biochem. Z.* **303**, 132.

CHAPTER IV

Flavoprotein enzymes

Discovery and properties of flavin

Warburg & Christian[1] in 1932 isolated from bottom yeast a yellow conjugated protein which they called the "yellow enzyme". By treatment with methyl alcohol they were able to resolve the enzyme into a colourless protein and a yellow prosthetic group. Shortly afterwards Kuhn[2] and his colleagues isolated lactoflavin, the substance responsible for the characteristic green fluorescence of whey. The properties of this milk compound were soon recognized to be practically identical with those of the prosthetic group of the Warburg-Christian "yellow enzyme". The Kuhn[3] and Karrer[4] schools independently determined the structure of lactoflavin and accomplished its synthesis.

Lactoflavin is 6 : 7-dimethyl-9-(*d*-1′ribityl)-*iso*alloxazin. The following method[3] is one of several available for the synthesis:

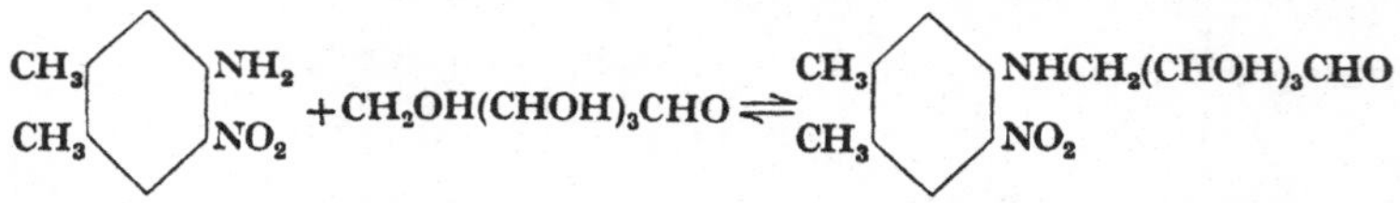

$$
\begin{array}{c}
CH_3 \\
CH_3
\end{array}
\bigcirc
\begin{array}{c}
NH_2 \\
NO_2
\end{array}
+ CH_2OH(CHOH)_3CHO \rightleftharpoons
\begin{array}{c}
CH_3 \\
CH_3
\end{array}
\bigcirc
\begin{array}{c}
NHCH_2(CHOH)_3CHO \\
NO_2
\end{array}
$$

1 : 2-dimethyl-4-nitro-   *d*-ribose   1 : 2-dimethyl-4-nitro-5-N
5-amino benzene   [*d*-1′ ribityl]

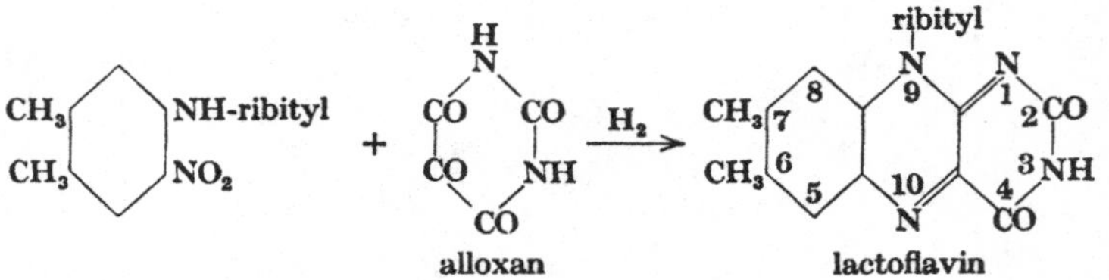

alloxan   lactoflavin

Lactoflavin shows in neutral solution an intense greenish fluorescence which disappears on addition of either strong alkali or acids. As little as $0\cdot05\gamma$ per ml. shows this fluorescence distinctly. The absorption spectrum is shown in Fig. 10.

On reduction with hyposulphite or Pt and H_2 the yellow colour is discharged, and leuco- or reduced flavin is formed. Two equivalents of H are involved in the reduction:

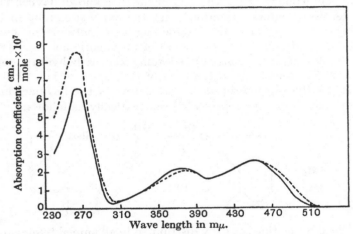

Shaking the solution of leuco-flavin with air restores the original colour. Leuco-flavin is autoxidizable, H_2O_2 being produced in the reaction with molecular oxygen. In acid solution the reduction of lactoflavin can be shown to proceed in two

Fig. 10. Absorption spectrum of riboflavin (———) and flavinadenine dinucleotide (- - - - - -). (Warburg & Christian [1938]. *Biochem. Z.* **298**, 150.)

stages[5], each of which involves one equivalent of H. The intermediary substance is red and there is good evidence that it is a semiquinone. For the theory of semiquinone formation and the structure of these free-radicle type of compounds the reader is referred to the reviews of Michaelis[6].

When irradiated in alkaline solution lactoflavin yields principally a photo-derivative known as lumiflavin[7] which has the constitution of 6 : 7 : 9-trimethyl*iso*alloxazine. The effect of light is to split off the four terminal carbon atoms of the ribityl side chain.

Phosphorylated flavin

Kuhn *et al.*[8] have been able to introduce a phosphoric acid group in position 5 of the ribityl side chain. The resulting synthetic lactoflavin-5-phosphate has been found to be identical with that of the prosthetic group of the Warburg-Christian flavoprotein of bottom yeast.

Straub[9] and Warburg & Christian[10] in 1938 discovered that the prosthetic group of the amino-acid oxidase was a flavin compound not identical with any of the then known flavins. The latter investigators established that the prosthetic group was a flavinadenine dinucleotide containing one molecule of both adenine and flavin, two molecules of phosphoric acid and two molecules of sugar. Strictly speaking the carbohydrate side chain of lactoflavin is in the form of the sugar alcohol, ribitol, and not the sugar *d*-ribose. The structure of the flavinadenine dinucleotide may be represented schematically as follows:

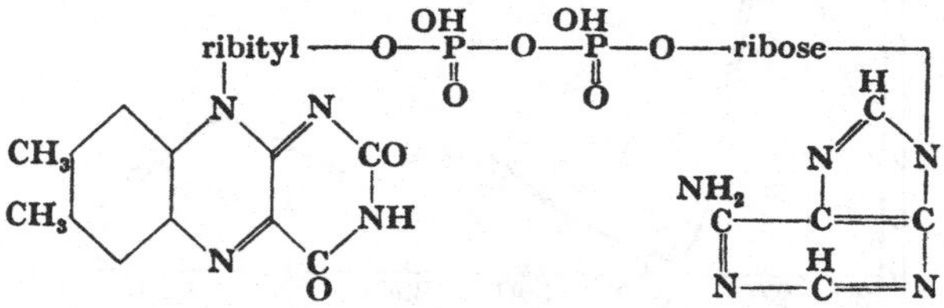

The dinucleotide is widely distributed in all animal tissues and in micro-organisms. Yeast offers the most convenient source of preparation. The following is a résumé of the essential stages in the method of Warburg & Christian[10]. The filtrate of a yeast suspension which has been heated to 75° is ⅔rds saturated with ammonium sulphate and extracted with phenol. The combined phenol extracts are mixed with ether and the dinucleotide is extracted with water. After removing ether by evacuation the aqueous solution is acidified with nitric acid to *ca.* pH 2 and the dinucleotide is precipitated as the silver salt. The precipitate is resuspended in water and decomposed with H_2S. The dinucleotide is almost completely adsorbed on the AgS precipitate and is eluted with dilute barium acetate. The combined eluates are mixed with ammonium acetate solution and are then concentrated *in vacuo* to dryness. The barium salt is twice as soluble at 60° as at room temperature. By taking advantage of this increase in solubility it is possible to separate the barium salt of the dinucleotide from the barium salts of contaminating adenine nucleotides. The yield of the pure flavinadenine dinucleotide obtained by this method is about 15 %.

The dinucleotide has the typical lemon-yellow colour of flavins and shows greenish fluorescence (cf. spectrum in Fig. 10). Hyposulphite reduces the dinucleotide to its leuco form, which in turn can be reoxidized by molecular oxygen. The dinucleotide is rather unstable. Biological activity is destroyed by short exposure to N alkali at room temperature or to N HCl at 100° (10). These treatments hydrolyse the dinucleotide into inactive fragments.

Properties of flavoprotein enzymes

Thus far seven flavoprotein enzymes have been isolated. Their properties and characteristics are summarized in Table IV. Four of these flavoproteins can catalyse the oxidation of the pyridine

Table IV

Source of flavo-protein	Prosthetic group	Position of absorption bands, mμ.	Substrate	Autoxidizability in air under physiological conditions
Yeast I	riboflavinphosphate	275, 380, 465	dihydrocoenzymes I and II	+
Yeast II	flavinadenine dinucleotide	275, 377, 455	,,	−
Heart	,,	274, 359, 451	,,	−
Kidney	,,		d-amino-acids	+
Milk	flavinadenine dinucleotide plus some unknown grouping or groupings	279, 360, 450	dihydrocoenzyme I aldehydes hypoxanthine	+ (hypoxanthine) − (dihydrocoenzyme I, aldehyde)
Liver	,,		aldehydes	+

nucleotides. This property may be fortuitous in the same sense
that all Fe porphyrin proteins show slight catalatic or peroxidatic properties. Straub's heart flavoprotein is unquestionably
specific for the catalysis of the oxidation of coenzyme I, being
some hundred times more active than the Warburg-Christian
flavoprotein and about seventeen times more active than milk
flavoprotein. Very likely experiment will disclose a more specific
catalytic role for the Warburg-Christian flavoprotein.

The prosthetic group of the Haas, Straub, and amino-acid
oxidase flavoproteins is flavinadenine dinucleotide; that of the
Warburg-Christian flavoprotein is flavin mononucleotide, i.e.
lactoflavinphosphate; and that of the milk and liver flavoproteins is flavinadenine dinucleotide in combination with one or
more groups of unknown constitution.

The Warburg-Christian flavoprotein in acid solution can be
resolved reversibly into protein and lactoflavinphosphate [11, 12].
By recombining the two portions in neutral solution, the original
conjugated flavoprotein is regenerated. If flavinadenine dinucleotide is used instead of the mononucleotide, a synthetic
flavoprotein [12] is formed, the properties of which are not
identical with any of the other natural flavoproteins. Thus the
flavoprotein of Haas is not simply the protein of the Warburg-
Christian flavoprotein combined with the dinucleotide. By using
massive concentrations of lactoflavin it is possible to obtain a
lactoflavin flavoprotein which exhibits to a slight degree the
catalytic properties of the lactoflavinphosphate flavoprotein [13].
In reconstituting the amino-acid oxidase from the protein and
flavin moieties the dinucleotide cannot be replaced by lactoflavinphosphate or by lactoflavin [9]. Specificity for the flavin is
sharper in this instance. The prosthetic group of milk flavo-

Table V. *Interchangeability of flavins*

Source of protein	Riboflavin	Riboflavin phosphate	Flavin-adenine dinucleotide
Yeast I	+	+	+
Yeast II	−	−	+
Amino-acid oxidase (kidney)	−	−	+
Xanthine oxidase (milk)	−	−	−
Fumaric hydrase (yeast)	+	+	+

protein can fulfil the role of prosthetic group for the amino-acid oxidase (14, 15). The reverse substitution, however, does not work. The various permutations and combinations are summarized in Table V.

Flavoproteins in the course of their catalytic function undergo a cycle of reduction and oxidation. For example, the Warburg-Christian flavoprotein is reduced by reduced coenzymes I and II and the leuco form is oxidized by molecular oxygen. The leuco forms of the Haas and Straub flavoproteins are for all practical purposes non-autoxidizable. Some autoxidizable "carrier" similar in behaviour to methylene blue has to be interposed between these leuco flavoproteins and molecular oxygen. The distinction between the autoxidizable and non-autoxidizable flavoproteins is made clear by the equations for cycles A and B:

A   reduced coenzyme I + flavoprotein → coenzyme I + leuco flavoprotein
    leuco flavoprotein + O_2 → flavoprotein + H_2O_2

B   reduced coenzyme I + flavoprotein → coenzyme I + leuco flavoprotein
    leuco flavoprotein + carrier → flavoprotein + reduced carrier
    reduced carrier + O_2 → carrier + H_2O_2

In reconstructed systems methylene blue and similar dyes serve as the intermediary carrier. It is not known which substances in the cell normally assume that role. There is some evidence that cytochromes a and b are concerned.

Warburg-Christian flavoprotein (yeast)

The preparation of this flavoprotein from bottom yeast* (the only source yet known) is as follows (1):

Maceration juice is treated with excess basic lead acetate solution and the precipitate is discarded. Excess lead is removed with phosphate. The pale yellow solution is mixed with ½ vol. of acetone at 0°. The precipitate is discarded. The same amount of acetone is again added and CO_2 is bubbled through the iced solution. A yellow oily precipitate forms which is separated by

---

* The yeast was obtained from the Patzenhofer brewery in Berlin. It is a special strain of bottom yeast.

decantation. The precipitate is redissolved and the acetone-CO_2 procedure is repeated. A crude preparation of flavoprotein is thus obtained in a yield of about 14 g. per kg. of dried yeast. The final stages in the purification which lead to the homogeneous material have been worked out by Theorell[11] and Kuhn & Desnuelle[16]. Cataphoresis at pH 4·6 was found by Theorell to separate the flavoprotein from contaminating polysaccharides as well as from inactive proteins[11]. This procedure is then followed by fractional $(NH_4)_2SO_4$ precipitation.

The molecular weight has been calculated from ultracentrifuge measurements to be *ca.* 70,000. Since the homogeneous preparation contains 0·66 % lactoflavinphosphate there can be only one molecule of prosthetic flavin per molecule of flavoprotein.

Treatment of powdered flavoprotein with 75 % methanol[1] irreversibly splits the pigment into insoluble denatured protein and soluble prosthetic group. There are two methods available for the reversible resolution of the Warburg-Christian flavoprotein. Theorell[11] dialyses a solution of flavoprotein against 0·02 N HCl at 0° for 72 hr. The prosthetic group is completely removed at the end of this time. The colourless contents of the dialysis sac are then dialysed against water to remove HCl. A clear solution of the protein component is thus obtained which can combine with flavinphosphate to form the original flavoprotein, cf. Fig. 11. Warburg & Christian[12] have recently worked out a simpler and more rapid method. A solution of flavoprotein is made 42 % saturated with respect to ammonium sulphate and acidified to pH 2·8 at 0°. The precipitate of protein is washed with half-saturated ammonium sulphate and redissolved in phosphate buffer at 0°. The yield of active flavin-free protein is *ca.* 78 %.

Flavinphosphate has a high affinity for the specific protein. At neutral pH no free flavinphosphate can exist in presence of a slight excess of the protein. Kuhn[13] attributes this high affinity to the fact that the protein is combined with lactoflavinphosphate via the phosphoric acid of the ribityl side chain and via the N in position 3 of the alloxan ring. The complex of specific protein and lactoflavin is linked only through the N of the alloxan ring and is therefore more easily dissociable. 6 : 7-Dimethyl-9-l-

araboflavin forms a similar type of dissociable flavoprotein, whereas the d-isomeride does not[13].

A mixture of the hexosemonophosphoric enzyme, hexosemonophosphate and coenzyme II does not react with molecular oxygen. On addition of the Warburg-Christian flavoprotein a reaction

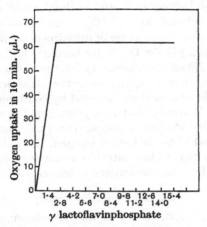

Fig. 11. Synthesis of Warburg-Christian flavoprotein. Combination of a given amount of protein with varying amounts of prosthetic group to form the original catalytically active flavoprotein. (Theorell [1935]. *Biochem. Z.* 278, 283.)

takes place[1]. Reduced coenzyme II reduces flavoprotein, and in turn reduced flavoprotein is oxidized by molecular oxygen:

hexosemonophosphate + coenzyme II $\xrightarrow{\text{enzyme}}$ phosphohexonate
+ reduced coenzyme II

reduced coenzyme II + flavoprotein → coenzyme II
+ reduced flavoprotein

reduced flavoprotein + O_2 → flavoprotein + H_2O_2

Thus flavoprotein acts as a catalytic link between reduced coenzyme II and molecular oxygen by undergoing a cycle of reduction and oxidation.

The above reconstruction was carried out manometrically with pure oxygen in the gas space. In air the reaction is much slower. The rate of oxidation of reduced flavoprotein by molecular oxygen depends on the partial pressure of oxygen. At the low oxygen tensions which exist, e.g. in animal tissues, flavoprotein would hardly be autoxidizable. It is only in artificial systems with unphysiological oxygen tensions that the Warburg-Christian flavoprotein can play the role of a link between reduced coenzyme II and oxygen. Under optimal conditions the flavoprotein is reduced and oxidized fifty times per min.

In an equimolecular mixture of dihydrocoenzyme I and flavoprotein at 0° and pH 7·8 the latter becomes half reduced in *ca.* 15 sec. With dihydrocoenzyme II as the reducing agent the velocity of reduction is slightly more than tenfold.

The synthetic flavoprotein obtained by combining the protein moiety of the Warburg-Christian yeast compound with flavin-adenine dinucleotide can catalyse the oxidation of dihydro-coenzymes I and II by molecular oxygen. The catalytic activity is only 70 % of that of the naturally occurring compound as far as reaction with molecular oxygen is concerned.

Haas flavoprotein (yeast)

Shortly after the discovery of the flavinadenine dinucleotide in animal tissues and micro-organisms, Haas(18) in Warburg's institute described the isolation from bottom yeast of a flavoprotein containing the dinucleotide as the prosthetic group. Maceration juice of "Schultzheiss-Patzenhofer" bottom yeast is acidified to pH 4·5 and made 60 % saturated with respect to $(NH_4)_2SO_4$. The precipitate is redissolved in water and the procedure repeated. After dialysis the solution is made 20 % with respect to ethyl alcohol at pH 4·5. After 18 hr. at 0° the precipitate containing the flavoprotein is centrifuged off and redissolved in water. The solution is brought to pH 10 and made 8 % saturated with respect to $(NH_4)_2SO_4$. The temperature is raised to $55\frac{1}{2}$° for 5 min. and then quickly lowered with ice. On neutralization to pH 7·4 a large precipitate of denatured protein is removed. Then follows fractional precipitation with $(NH_4)_2SO_4$ between the limits of 45–60 % saturation. The final procedure involves adsorption on "Fasertonerde" and elution with alkaline borate solution. The flavoprotein is then precipitated with 20 % alcohol at pH 4·6. The yield of flavoprotein calculated on the amount present in the maceration juice is 7 %. The Warburg-Christian flavoprotein is removed completely after the first two stages in the isolation.

The best preparations contain 0·7 % flavinphosphate. As yet there is no ultracentrifugal or cataphoretic data as to the homogeneity of the 0·7 % flavinphosphate preparation. If it is homogeneous, then there must be one molecule of flavinadenine dinucleotide combined with one molecule of protein of molecular weight *ca.* 70,000.

The flavoprotein shows three absorption bands with peaks at 275, 377 and 455 mμ. respectively. Concentrated solutions are more greenish in appearance than those of the Warburg-Christian flavoprotein. Comparison of the spectrum of the new flavoprotein with that of the constituent dinucleotide shows that the effect of the protein is to shift the bands towards the red end of the spectrum.

The catalytic properties of the Haas flavoprotein are qualitatively but not quantitatively the same as those of the Warburg-Christian compound. The following are the respective velocity constants:

	$k_{oxid.}$	$k_{red.}$	Ratio $\dfrac{k_{oxid.}}{k_{red.}}$
Warburg-Christian	10×10^4	6×10^6	0·0167
Haas	$1·4 \times 10^4$	22×10^6	0·00064

The dimensions of k are $\dfrac{1}{\text{mols per litre} \times \text{min.}}$. $k_{oxid.}$ is the velocity constant for the rate of oxidation of reduced flavoprotein by molecular oxygen and $k_{red.}$ is the velocity constant for the rate of reduction of flavoprotein by dihydrocoenzyme II. At physiological tensions of oxygen the Haas flavoprotein is virtually non-autoxidizable and in reconstructed systems it requires an intermediary carrier for its reaction with molecular oxygen. It is interesting that the Haas flavoprotein is more rapidly reduced by dihydrocoenzyme II than the Warburg-Christian compound.

In a solution 45 % saturated with respect to $(NH_4)_2SO_4$ and at pH 0·4 the Haas flavoprotein is split into protein and prosthetic groups, the former being precipitated. The precipitate is washed with 45 % saturated $(NH_4)_2SO_4$ at 0° and finally dissolved in water. The clear colourless solution contains the native protein which can combine with the dinucleotide to form the original conjugated compound. This resolution can only be carried out with fairly crude preparations of the flavoprotein. The high acidity necessary for the resolution completely denatures the highly purified flavoprotein. The specific protein cannot form a catalytically active complex with lactoflavinphosphate.

Amino-acid oxidase (kidney)

Properties of enzyme. Krebs[19] isolated from liver and kidney an enzyme which catalyses the oxidation of *d* or non-physiologically occurring α-amino acids. β-Hydroxyglutamic acid is the only member of the *d* series of α-amino acids which is not attacked. Sulphur-containing amino acids show the highest rate of oxidation (cf. Table VI). None of the *l* series (i.e. natural

Table VI. *Rate of oxidation of various amino acids in the presence of d-amino-acid oxidase*

(Krebs [1935]. *Biochem. J.* 29, 1620.)

Amino acid	Oxygen uptake in μl. per 20 min.	Amino acid	Oxygen uptake in μl. per 20 min.
dl-Methionine	269	dl-Tyrosine	51
d-isoLeucine	176	dl-Arginine	41
d(−)-Phenyl-alanine	143	dl-α-Phenyl-α-aminoacetic	35·5
dl-Alanine	138	dl-Tryptophan	19
dl-S-Ethylcysteine	131	d(+)-Histidine	16·5
dl-norLeucine	121	dl-Aspartic	15
d(−)-Valine	106	dl-α-Aminocaprylic	11
dl-Serine	78	dl-α-Amino-β-γ-di-hydroxybutyric	9
dl-Leucine	55		
dl-α-Aminobutyric	55	dl-Glutamic	5
dl-Cystine	54	dl-β-Hydroxy-glutamic	0
dl-norValine	53		

series) are oxidized. It is interesting that glycine, which has no asymmetric centre, is unattacked whereas phenylglycine, which is asymmetric, is oxidized. Amino acids with the amino group in the β position, e.g. β-alanine, β-aminobutyric acid, etc., and dipeptides such as alanylglycine and leucylglycine are inactive as substrates.

Oxygen can be replaced by other hydrogen acceptors only in the cases of a few amino acids, leucine particularly (Bernheim & Bernheim (45)). In those cases in which oxygen can be substituted by dyes, the rates with the latter are about fifty times slower than with oxygen.

The soluble amino-acid oxidase occurs in the liver, kidney and intestinal mucosa of all the mammals and birds which have thus far been tested. Kidney is about four times as active per unit tissue as liver. According to Krebs (19), there are two enzymes in liver and kidney which catalyse the oxidative deamination of amino acids: (1) the system described above and referred to as d-amino-acid oxidase, and (2) a system which cannot be separated from cells and which, like the respiration of the intact cell, is strongly inhibited by cyanide, octyl alcohol and drying. This system, called by Krebs l-amino-acid deaminase, oxidizes the natural amino acids.

It is significant that the oxidation of d-amino acids in tissue slices is partially inhibited by cyanide, although cyanide in the same concentration has no effect on the extracted enzyme. This discrepancy between *in vitro* and *in vivo* experiments is not unusual.

Table VII. *Effect of dilution on the rate of deamination of*
l(−)-aspartic acid in kidney brei

(Krebs [1935]. *Biochem. J.* **29**, 1620.)

Tissue	Temp. ° C.	Final dilution of tissue	Time in min.	Blank	With $l(-)$-aspartic acid	With $l(-)$-aspartic acid corrected for blank
					μl. ammonia formed	
Pig kidney	37·5	4-fold	50	82	330	248
,,	37·5	8-fold	50	55	67·5	12·5
Pig kidney	37·5	4-fold	30	76·5	127	50·5
,,	37·5	8-fold	30	42	74	32
,,	37·5	16-fold	30	41·5	53	11·5
Pig kidney	37·5	4-fold	40	119	375	257
,,	37·5	8-fold	40	58	70	12
Sheep kidney	37·5	4-fold	40	159	242	83
,,	37·5	8-fold	40	108	125	17

In comparing different dilutions of brei the same dry weight was used in each experiment, the variable being the volume of total suspension fluid.

The main difference between the d- and l-amino-acid oxidases appears to lie in their extractability from ground tissue. The former is easily extracted, whereas the latter (the glutamic enzyme excepted) cannot be separated from the insoluble particles of the cell. The difference is not merely one of solubility, since the l-amino-acid oxidase is found neither in the aqueous extract nor in the residual ground tissue. The l-oxidase or series of oxidases belongs to the category of cellular oxidations which are intimately bound up with the structure and organization of the cell. Some experiments of Krebs indicate that the inactivation of the l-amino-acid oxidase is accomplished neither by grinding nor by disruption of cell structure but by the dilution which necessarily accompanies extraction. The extent of the dilution determines the extent of the inactivation (cf. Table VII).

Keilin & Hartree[20] have established that the d-amino-acid system produces H_2O_2* in the reaction with molecular oxygen. By adding catalase and alcohol to the oxidase system they were able to demonstrate oxidation of alcohol to aldehyde by the H_2O_2 formed in the oxidative deamination.

N-Methylalanine is oxidized in presence of the enzyme to pyruvic acid and methylamine[20]. Similarly proline (secondary amine) can be oxidized to α-keto-δ-amino valeric acid[46]. N-Dimethylalanine and α-methylalanine are not oxidized. These facts support the imino acid theory for the oxidation of amino acids.

$$\begin{array}{l}R \\ >\!\!CHNH_2 + acceptor \rightarrow \\ COOH\end{array} \quad \begin{array}{l}R \\ >\!\!C=NH + reduced\ acceptor \\ COOH\end{array}$$

$$\begin{array}{l}R \\ >\!\!C=NH + H_2O \rightarrow \\ COOH\end{array} \quad \begin{array}{l}R \\ >\!\!C=O + NH_3 \\ COOH\end{array}$$

The primary step in deamination is the removal of two hydrogen atoms and the formation of the unsaturated imino acid. The oxidase is involved in this reaction. The hydrolysis of the imino acid to the keto acid and NH_3 is presumably uncatalysed. The theory explains why N-methylalanine is attacked, whereas

---

* In highly purified preparations of the amino-acid oxidase devoid of catalase, H_2O_2 produced in the reaction oxidizes pyruvic acid to acetic acid and CO_2[21]. This oxidation is spontaneous and does not require a catalyst.

N-dimethyl is not. In the latter case there is no possibility of imino-acid formation.

Resolution of the enzyme. The amino-acid oxidase is easily split into its components by the following modification of Straub's method (22). The aqueous extract of the acetone powder of pig's kidney is acidified to pH 4·5 at 0° and made 33 % saturated with respect to $(NH_4)_2SO_4$. The precipitate is centrifuged and redissolved in water. The precipitation is carried out three times in all. The flavin-free protein solution is stable for weeks at 0°. Warburg & Christian (12) effect the resolution at pH 2·8, though with greater loss of activity due to denaturation of the protein.

Negelein & Brömel (21) have effected a high degree of purification of the specific protein of the amino-acid oxidase. They estimate the purity level of their best preparation to be *ca.* 70 %. Their method consists essentially in warming the slightly acidified solution of the protein to 38°, thereby denaturing impurities, and in repeated precipitations of the protein with 20 % saturation of $(NH_4)_2SO_4$ at pH 5·0. 1 mg. of their purest preparation in presence of excess dinucleotide and alanine takes up 26,000 μl. O_2 per hr. at 37°. With excess of protein and alanine, dinucleotide equivalent to 1γ of the barium salt catalyses the uptake of 2100 μl. O_2 per hr. at 37°.

The amino-acid oxidase dissociates at pH 8·5 to the extent of some 90 %. The experimental evidence for this dissociation is derived by Warburg & Christian (10) from the following kinetic consideration. Let C represent the concentration of dinucleotide which in presence of the other components of the system gives rise to velocity of oxidation V, and C_∞ the concentration of dinucleotide which gives rise to the maximum velocity V_E. If we assume that the free protein combines with dinucleotide to form a dissociating compound, the equilibrium constant K for this reaction is equal to $C \times \dfrac{\text{[free protein]}}{\text{[bound protein]}}$. It can easily be seen that the ratio $\dfrac{\text{[free protein]}}{\text{[bound protein]}}$ is equal to the ratio $\dfrac{V_E - V}{V}$. Table VIII summarizes the data of an experiment designed to test the validity of this relationship. The value of K is constant over a wide range of concentration of the flavinadenine dinucleotide. The dimensions of K are γ barium salt per ml. The K value means that half the total available protein is bound when the concentration of the dinucleotide is $0\cdot 23\gamma$ barium salt per ml. or $0\cdot 196\gamma$ free dinucleotide per ml. In terms of molarity $K = 2\cdot 5 \times 10^{-7}$ mols. per litre. Assuming that the molecular

Table VIII. *The constancy of K with varying amounts of prosthetic group*

(Warburg & Christian [1938]. *Biochem. Ž.* 298, 150.)

K	μl. O_2 per 10 min.	μg. barium salt per 2·4 ml.
—	0	0
0·228	6·3	0·05
0·224	16·4	0·15
0·228	35·8	0·50
0·258	53·2	1·5
0·216	67·4	4·5
—	75·1 (V_E)	18·0

Conditions: 20γ protein, 0·2 ml. 4·5 % *dl*-alanine, 1 ml. $M/10$ pyrophosphate, pH 8·3, final volume 2·4 ml.

weight of the protein is *ca.* 65,000, and that one molecule of protein combines with one molecule of dinucleotide (mol. wt. 926), it follows that 29γ of 70 % pure protein should be equivalent to $0\cdot 40\gamma$ of the barium salt of the dinucleotide. The limiting velocity is approached with $4\cdot 5\gamma$ barium salt. Hence the degree of dissociation when the limiting velocity is approached is $\left(1 - \dfrac{0\cdot 40}{4\cdot 5}\right) \times 100$ or *ca.* 91 %. If the above assumption as to the molecular weight of the protein and the percentage of dinucleotide is correct, we may consider that the amino-acid oxidase as prepared from fresh kidney is largely dissociated into its component parts. Although the amino-acid oxidase behaves as a dissociated flavoprotein in solution, it precipitates out of solution at neutral pH as a conjugated flavoprotein. There is evidence that the reduced form of the oxidase does not dissociate. In that respect the amino-acid oxidase differs from the diphospho- and triphosphopyridinoprotein enzymes.

It is interesting to compare the relative degrees of dissociation of various conjugated proteins[21]. The concentration of prosthetic group at which half the available protein is bound provides a measure of the degree of the dissociation.

	Concentration of prosthetic group for half-saturation (mols/litre)
Warburg-Christian flavoprotein	Very small
Haas flavoprotein	,,
Milk flavoprotein	,,
Heart flavoprotein	,,
Amino-acid oxidase flavoprotein	0.025×10^{-5}
Hexosemonophosphate pyridinoprotein	1×10^{-5}
Alcohol pyridinoprotein	3.2×10^{-5}
Triosephosphoric pyridinoprotein	3.2×10^{-5}

The amino-acid oxidase is intermediary between the rigidly bound flavoproteins and the largely dissociated pyridinoproteins.

Straub flavoprotein (heart)

The elucidation of the mechanism whereby the dihydro-pyridine nucleotides are oxidized in animal tissues is the result of comparatively recent work. For some time it was gratuitously assumed that animal tissues contain the Warburg-Christian flavoprotein, although no one was able to demonstrate its presence. Later an enzyme was discovered (23, 24) in all animal tissues which specifically catalyses the oxidation of the dihydro-pyridine nucleotides. The crude preparations of this enzyme (known as coenzyme factor or diaphorase) were found to be more active per mg. dry weight than the pure Warburg-Christian flavoprotein (23). Coenzyme factor occurs in association with insoluble particles. This association made difficult the task of its purification and isolation. The chemical nature of the coenzyme factor was obscure until Straub (25) succeeded in separating a flavoprotein from the particles with which the coenzyme factor is found associated. The soluble flavoprotein has all the catalytic properties of the coenzyme factor (22). Heart flavoprotein and coenzyme factor can be considered to be the same compound in two different physical states.

The following is a résumé of the method of Straub (25) for preparing the flavoprotein of pig heart. The mince of heart muscle is thoroughly washed with tap water and then ground to a fine paste with sand and alkaline phosphate buffer. The mixture is centrifuged and the supernatant fluid is acidified to pH 4·6. The precipitate is resuspended in water. This suspension represents the usual preparation of the coenzyme factor. The insoluble particles can be washed several times with buffer solutions or with water without appreciable loss in activity. But if the factor preparation is made 2 % with respect to ammonium sulphate and 3 % with respect to ethyl alcohol, and kept at 43° for 10–15 min., some 14 % of the enzyme activity can be separated from the particles. After centrifuging off the insoluble material the supernatant yellow fluid containing the flavoprotein is mixed with alumina $C\gamma$. The flavoprotein is then eluted from the adsorbent with alkaline phosphate buffer. For the final purification the following procedures are employed. The flavoprotein is precipitated between the limits of 45 to 60 % saturation of ammonium sulphate. Two such fractionations purify the enzyme some ninefold. An additional method of purification consists in heating the flavoprotein solution to 60° for 5 min. This procedure denatures some of the protein impurities and leaves intact the flavoprotein. The yield of Straub's method is $ca.$ 100 mg. of flavoprotein from 1 kg. dry weight of heart.

The best preparations of heart flavoprotein contain 0·65 % flavinphosphate. Ultracentrifugal measurements of Philpot indicate a molecular weight of $ca.$ 70,000 for the protein.

Heart flavoprotein is lemon yellow in colour and shows in solution an intense greenish fluorescence[25]. It is the only representative of the class of flavoproteins which is known to fluoresce like free flavins. The absorption spectrum has three bands with maxima at 274, 359 and 451 mμ. respectively. The 451 band is complex, and when observed in the microspectroscope shows a second component with a peak at about 490 mμ. Recently Keilin & Smith[26] have called attention to the fact that the faint bands at 450 and 490 mμ. observed in the field of spectroscopes of small dispersion are due to some pigment (presumably flavoprotein) present in the retina of the human eye.

The catalytic activity of heart flavoprotein is indeed formidable. Under optimum conditions each molecule of the enzyme catalyses the oxidation of $ca.$ 8000 molecules of dihydro-

coenzyme I per minute by carriers such as methylene blue[22].
Fig. 12 shows the dependence of the turnover number on the
concentration of flavoprotein. Under optimum conditions 1γ of
the flavoprotein (or $0\cdot0065\gamma$ of flavinphosphate equivalent) cata-

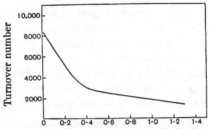

Concentration of flavoprotein in μg. flavinphosphate

Fig. 12. Turnover number of heart flavoprotein. (Corran, Green & Straub
[1939]. *Biochem. J.* 33, 793.)

lyses the absorption of 180 μl. O_2 per hr. in the lactic enzyme
system with methylene blue as the "carrier" between dihydrocoenzyme I and molecular oxygen. Heart flavoprotein also
catalyses the oxidation of dihydrocoenzyme II. This catalysis has
not as yet been studied quantitatively.

The yellow-green colour of a heart flavoprotein solution is
rapidly bleached on addition of dihydrocoenzyme I[22]. The
greenish fluorescence disappears simultaneously. The leuco or
reduced form of heart flavoprotein is comparatively sluggishly
oxidized by molecular oxygen in air, though it is rapidly oxidized

Table IX. *Catalytic effect of flavoprotein in the lactic
enzyme system*

(Corran, Green & Straub [1939]. *Biochem. J.* 33, 793.)

Lactic enzyme (ml.)	1·5	1·5	1·5	1·5	1·5
M lactate (ml.)	0·2	—	0·2	0·2	0·2
0·075 % coenzyme I (ml.)	1·0	1·0	—	1·0	1·0
2 M HCN (ml.)	0·2	0·2	0·2	0·2	0·2
0·5 % methylene blue (ml.)	0·2	0·2	0·2	—	0·2
Flavoprotein (ml.) (17·5 μg. per ml. as flavinphosphate)	0·1	0·1	0·1	0·1	—
Water	—	0·2	1·0	0·2	0·1
μl. O_2 in 5 min.	340	14	14	14	14

by methylene blue, pyocyanine, etc. The mechanism of heart flavoprotein catalysis involves a cycle of reduction by dihydrocoenzyme I and oxidation by carriers. In reconstructed systems with catalytic amounts of flavoprotein (cf. Table IX) no appreciable reaction with molecular oxygen is observed in absence of methylene blue. We may regard the slow autoxidation of leuco-flavoprotein as having no physiological importance. In animal tissues there is presumably one or more substances which fulfil the role of methylene blue.

The aerobic oxidation in animal tissues of metabolites such as lactate, malate, triosephosphate, β-hydroxybutyrate, triose, etc. can be resolved into at least four component reactions:

(1) substrate + coenzyme I → oxidized substrate
                                    + dihydrocoenzyme I

(2) dihydrocoenzyme I + flavoprotein → coenzyme I
                                    + leuco-flavoprotein

(3) leuco-flavoprotein + carrier → flavoprotein + reduced carrier

(4) reduced carrier + O_2 → carrier

The nature of the physiological carrier is the outstanding deficiency in our knowledge of the aerobic oxidation of metabolites via coenzyme I.

In tissue slices and in certain reconstructed systems the oxidation of lactate, malate, β-hydroxybutyrate, triosephosphate, etc. is found to be cyanide sensitive. Since the oxidation of the metabolites by coenzyme I and the reduction of flavoprotein by dihydrocoenzyme I are cyanide insensitive reactions the further reaction between leuco-flavoprotein and molecular oxygen via the carrier must be cyanide sensitive. The available evidence is in favour of the view that the cytochrome oxidase is the cyanide-sensitive enzyme(23). Presumably leuco-flavoprotein is oxidized by certain of the cytochrome components and the cytochrome oxidase catalyses the oxidation of the ferro cytochromes by molecular oxygen. The standard cytochrome oxidase preparation of heart contains, among other substances, the three cytochrome components (mostly a and b), cytochrome oxidase, flavoprotein and several pyridinoprotein enzymes, e.g. lactic, malic, etc. On addition of lactate or malate and coenzyme I under anaerobic conditions, cytochromes a and b are reduced at once (23). Whether reduced flavoprotein reacts directly with these cytochrome components, or whether there are intermediary stages, is as yet an unsolved problem. Haas *et al.* (47) have reported the isolation of a flavoprotein from yeast which catalyses the oxidation of dihydrocoenzyme II by cytochrome c. It is possible that special flavoproteins may be concerned in the reaction between the dihydropyridine nucleotides and the cytochromes.

The prosthetic group of heart flavoprotein is flavinadenine dinucleotide (25). Boiling the solution of the flavoprotein liberates the prosthetic group from the denatured protein. The liberated prosthetic group can then act as the coenzyme of the amino-acid oxidase. It is interesting that the flavoprotein itself has no activity in the amino-oxidase test.

Heart flavoprotein is remarkably stable to high temperatures. For example, exposure to 80° for some minutes does not destroy its catalytic activity. Yet when the flavoprotein is attached to the insoluble particles (i.e. in its natural physical state) exposure to 55° for 10 min. destroys practically all catalytic activity (22). In other respects the properties of soluble and bound flavoprotein are identical: e.g. both are rapidly destroyed in solution more acid than pH 4 and more alkaline than pH 9.

Coenzyme factor has been shown to be universally distributed in animal tissues, and enzymes similar in properties have been found in certain micro-organisms, e.g. *Bact. coli*, yeast, etc. (27). Since heart flavoprotein is identical with the coenzyme factor it must be universally distributed in animal tissues and either the same or a very similar flavoprotein must presumably account for the coenzyme factor activity of certain micro-organisms.

Euler and his colleagues (44) deduce from indirect evidence the existence in animal tissues of a flavoprotein which specifically catalyses the oxidation of dihydrocoenzyme II. This hypothetical flavoprotein is not considered to be identical with that of Straub. It is of interest to note that heart flavoprotein as isolated by Straub catalyses the oxidation of both dihydrocoenzyme I and II.

Fumaric hydrogenase

Fischer & Eysenbach[27] discovered in yeast an enzyme which catalyses the reduction of fumarate by the reduced forms of various oxidation-reduction dyes. Fumarate is reduced to succinate whilst the reduced dye is oxidized to the coloured form. The enzyme now known as fumaric hydrogenase was first observed in crude extracts of the Warburg-Christian flavoprotein. It was, however, soon demonstrated that fumaric hydrogenase was a flavoprotein not identical with the Warburg-Christian flavoprotein. Recently Fischer *et al.* [28] have succeeded in separating the two flavoproteins by cataphoresis in alkaline solution.

The prosthetic group of fumaric hydrogenase can be reversibly split off by precipitating the protein with ammonium sulphate in strongly acid solution[28]. On addition of flavin or flavin-phosphate to a solution of the protein separated from its prosthetic group only a small percentage of the original catalytic activity is restored. Flavinadenine dinucleotide, however, in minute amounts restores practically the original activity, e.g.

γ Dinucleotide	Time for reoxidation of reduced methyl viologen in min.
0·0004	120
0·0008	65
0·004	43
0·04	5·5
0·4	3·3
0·8	2·5
4·0	1·8
8·0	1·5

(Original activity 1·33 min.)

Combination of the protein and dinucleotide takes place fully only after several hours.

Prolonged dialysis of fumaric hydrogenase[28] leads to a loss in catalytic activity. This has been correlated with the loss of prosthetic group by dialysis. Presumably fumaric hydrogenase dissociates to some degree even in neutral or slightly acid solution.

Under optimum conditions[28] each molecule of the enzyme catalyses the reduction of 2500–3000 molecules of fumarate to succinate in 1 min. (T.N. = *ca.* 2750). From the point of view of catalytic efficiency fumaric hydrogenase is intermediate between heart flavoprotein (T.N. = 8000) and d-amino-acid oxidase (T.N. = 2000).

The following mechanism of the catalytic reduction of fumarate seems most probable from the available facts:

reduced dye + flavoprotein → dye + reduced flavoprotein

reduced flavoprotein + fumarate → flavoprotein + succinate

Flavoprotein undergoes a cycle of reduction by the leuco dye and oxidation by fumarate. There is no specificity shown by flavoprotein towards the reducing agent. A large number of reduced oxidation-reduction indicators have been found effective. Hyposulphite can replace the reduced dyes as reductant of the flavoprotein. There are no data as to the nature of the physiological reducing agent. Possibly the flavoprotein responsible for the reduction of fumarate may also be reducible by the dihydropyridine nucleotides. In that event fumaric hydrogenase may act as a link between the pyridinoprotein enzyme systems and the systems concerned in the metabolism of the C_4 dicarboxylic acids.

Fumaric hydrogenase is reported [27] to catalyse only the reduction of fumarate to succinate and not the oxidation of succinate to fumarate. It should be pointed out, however, that it is theoretically impossible for an equilibrium reaction to be catalysed in one direction only.

The following substances can replace fumarate in the catalytic system [27]: maleate, crotyl alcohol, phenylcrotyl alcohol and geraniol. They have, however, less than $\frac{1}{50}$th the efficiency of equivalent amounts of fumarate. Muconate, itaconicate, crotonate and phenylcrotonate are inactive as oxidants.

Fumaric hydrogenase [27] is inactivated above 60°. It is most active catalytically in the pH range 6–7, being destroyed below pH 4·5 and above pH 10.

Milk flavoprotein (xanthine, aldehyde, dihydrocoenzyme I oxidase)

Properties of the enzyme. Xanthine oxidase was first discovered in milk by Hopkins et al. [29] and has been more thoroughly studied than any other oxidizing enzyme. Hopkins and Dixon have both made classical contributions to our knowledge of the properties and kinetics of the oxidase.

The xanthine oxidase of milk and liver can catalyse the oxidation of aldehydes as well as of xanthine. It has long been assumed that an enzyme distinct from the xanthine oxidase was responsible for the activation of aldehydes. But the recent study of Booth (31) has confirmed the earlier view of Dixon (32) that the xanthine oxidase catalyses both oxidations (cf. review of Dixon (42)). The aldehyde activating enzyme was originally known as the "Schardinger" enzyme after its discoverer (30) before identity with the xanthine oxidase was established.

The xanthine oxidase shows a considerable degree of specificity for the purines attacked. The following table, taken from Dixon (42), gives the relative velocities with which different purines are oxidized:

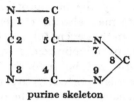

purine skeleton

Purine	Relative rate	Reference no.
6-Oxy-*P*	200	(29)
2 : 6-Dioxy-*P*	100	(29)
2-Thio-6-oxy-*P*	64	(33)
6 : 8-Dioxy-*P*	100	(33)
8-Oxy-*P*	12	(35)
2 : 8-Dioxy-*P*	1 ?	(35)
6-Amino-*P*	2	(32)
6-Amino-2-oxy-*P*	8	(35)
6-Amino-8-oxy-*P*	12	(35)

(*P* = purine)

No other purines of a long list have been found to be active. There is no specificity shown towards aldehydes (31). Any nontoxic aldehyde, aromatic or aliphatic, can be oxidized in presence of the enzyme.

There is an interesting difference between purine and aldehyde activation. Whereas the rate of oxidation of purines is maximal at low concentrations ($0.00015\ M$) (34), maximal rates with aldehydes are reached only at relatively high concentrations (*ca.*

0·06 M). This must mean that the affinity of the enzyme for purines is about 400 times greater than for aldehydes.

A satisfactory explanation of how one enzyme can catalyse the oxidation of two classes of substances as widely different as aldehydes and purines has yet to be given. It is difficult to imagine that the active group (as distinct from the prosthetic group) of the xanthine oxidase should be capable of a dual activating role. Yet the experimental evidence is clear that aldehyde specifically inhibits the oxidation of purine and vice versa. The interpretation of this competitive inhibition phenomenon is either that the same active group is concerned in the activation of both groups of substrates or that there are two active groups in such close juxtaposition that activity at the one interferes with activity at the other.

Hypoxanthine and xanthine are oxidized to uric acid in presence of the oxidase, whilst oxygen is reduced to H_2O_2 (Hopkins[29], Dixon[36]). The production of H_2O_2 allows of coupled oxidations, e.g. if nitrite and peroxidase are added to the xanthine oxidase system nitrate and uric acid are formed (Harrison[37]). Similarly, if catalase and ethyl alcohol are added acetaldehyde and uric acid are formed[20].

$$\text{hypoxanthine} + O_2 \xrightarrow{\text{xanthine}\atop\text{oxidase}} \text{xanthine} + H_2O_2$$

$$\text{xanthine} + O_2 \xrightarrow{\text{xanthine}\atop\text{oxidase}} \text{uric acid} + H_2O_2$$

$$H_2O_2 + \text{nitrite} \xrightarrow{\text{peroxidase}} \text{nitrate} + H_2O$$

$$H_2O_2 + \text{alcohol} \xrightarrow{\text{catalase}} \text{aldehyde} + H_2O$$

Coupled oxidations refer to oxidations in one system which depend upon a substance produced by oxidation in a second system.

Xanthine oxidase can catalyse the anaerobic dismutation of xanthine to hypoxanthine and uric acid[38,39]:

$$2 \text{ xanthine} \rightleftharpoons \text{hypoxanthine} + \text{uric acid}$$

Starting with hypoxanthine and uric acid, xanthine will be formed until equilibrium is reached. At neutral pH the equilibrium ratio hypoxanthine : xanthine : uric acid is $0·68 : 1 : 0·68$[38].

In the dismutation one molecule of xanthine is reduced to hypo-xanthine and a second is oxidized to uric acid. The formulae for these three purines are given below:

hypoxanthine   xanthine   uric acid

Uric acid reacts with aldehydes, particularly salicylaldehyde, in presence of the xanthine oxidase (31):

uric acid + salicylaldehyde → hypoxanthine + salicyclic acid

The xanthine oxidase by virtue of its ability to activate both purines and aldehydes is capable of catalysing the oxidation of aldehyde by uric acid without the intervention of an external hydrogen carrier. This points to some oxidizable and reducible group within the enzyme molecule itself.

Isolation of enzyme and nature of prosthetic group. Ball (15)* in 1938 described the isolation from cow's milk of a flavoprotein which showed all the properties of the xanthinealdehyde oxidase. He made out a convincing case that the flavoprotein was in fact identical with the enzyme. At the same time Corran & Green (14) isolated from cow's milk a flavoprotein devoid of xanthine oxidase activity but capable of catalysing the oxidation of dihydro-coenzyme I. The properties of the flavoprotein isolated in these two investigations, from their description, appeared identical except with respect to catalytic activity.

Corran *et al.* (40) have cleared up the apparent discrepancy. Milk flavoprotein has been shown to be an enzyme which cata-lyses the oxidation of three substrates, viz. hypoxanthine, alde-hyde and dihydrocoenzyme I. By appropriate treatment it is possible to eliminate hypoxanthine and aldehyde activity without impairing activity towards dihydrocoenzyme I. No method has

* Dixon & Kodama (48) in 1926 had obtained a preparation of the xanthine oxidase not far removed from "pure", although unable at that time to correlate the activity with the colour.

yet been found either for eliminating dihydrocoenzyme activity without affecting the other two or for eliminating hypoxanthine activity leaving aldehyde activity unaffected (and vice versa).

The following is the method of preparation used by Corran *et al.* (40). Whole fresh cow's milk is saturated with NaCl at 30°. The clear filtrate is half saturated with $(NH_4)_2SO_4$. The precipitate is dissolved in buffer and the solution made 40 % saturated with respect to $(NH_4)_2SO_4$. The precipitate is dissolved in water and the solution is dialysed. Then follows precipitation at 0° and *p*H 4·6 with 13 % alcohol. The precipitate is dissolved in water and the enzyme is adsorbed on alumina *cy* and eluted with $M/2$ phosphate buffer *p*H 7·2. For final purification several fractional $(NH_4)_2SO_4$ precipitations are carried out between the limits of 35–45 % of saturation. The best preparation obtained is only *ca.* 70 % homogeneous, though all the colour is associated with one protein component (49). The yield of flavoprotein at this level of purity is *ca.* 2 %. The catalytic activity per mg. dry weight is *ca.* 1000 times greater than that of milk.

A dilute solution of milk flavoprotein is brown in colour; more concentrated solutions are brownish red. The characteristic of all flavin compounds, whether free or combined with protein, is a true yellow colour. It is therefore at once apparent that some coloured substance other than flavin must be combined with the protein of milk flavoprotein. Flavin compounds free or combined are completely bleached by hyposulphite. Milk flavoprotein is bleached by hyposulphite only to the extent of 60 %, leaving a brownish residual colour. Furthermore, only 35 % of the total absorption of light by a solution of milk flavoprotein at 450 mμ. (main absorption band) can be accounted for by flavin. All the evidence therefore points to the presence of some other coloured grouping in the molecule in addition to flavin.

Fig. 13 represents the absorption spectrum of milk flavoprotein. There are three bands with maxima at 452, 360 and 280 mμ. respectively. On reduction with hyposulphite the typical flavin bands with peaks at 452 and 360 mμ. disappear, leaving a high residual general absorption.

Acidification of a solution of milk flavoprotein denatures the protein and liberates the prosthetic flavin. This, unlike the parent substance, has a pure yellow colour with a spectrum indistinguishable from that of flavinadenine dinucleotide. Chemical analysis has shown that each mole of the liberated flavin contains one mole adenine and two moles phosphoric acid. In agreement with this correspondence in chemical constitution is the fact that the flavin group of milk flavoprotein can replace flavinadenine dinucleotide in the d-amino-acid oxidase test system.

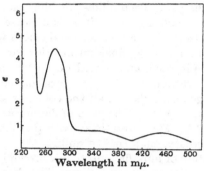

Fig. 13. Absorption spectrum of oxidized milk flavoprotein. (Corran, Dewan, Gordan & Green [1939]. *Biochem. J.* 33, 1694.)

Corran *et al.* [40] find that their preparation which is *ca.* 70 % homogeneous contains 0·27 % flavinphosphate. The homogeneous preparation therefore should contain 0·38 % flavinphosphate. Since the molecular weight of milk flavoprotein is *ca.* 280,000 (Philpot [41]) it follows that there are probably two flavin groups per molecule of flavoprotein. The final decision on the number of flavin groups must await the isolation of the homogeneous compound.

The flavoprotein as prepared above catalyses the oxidation of hypoxanthine, aldehyde and dihydrocoenzyme I. It is possible by a variety of treatments to inactivate the enzyme with respect to hypoxanthine and aldehyde activities without impairing the third function. The most effective treatments are (1) drying the enzyme and (2) incubation with $M/10$ HCN (cf. Dixon & Keilin [43]). These treatments have no observable effect on the flavoprotein, either chemically or spectroscopically. That any change has taken place is shown only in the catalytic tests. These differential inactivation experiments prove conclusively that the active group concerned in the oxidation of hypoxanthine and aldehyde is not identical with that concerned in the oxidation of dihydrocoenzyme I. The question then arises whether the two respective active groups are associated with the same molecule. The experimental evidence is wholly in favour of the view that all three catalytic activities are functions of the same flavoprotein.

Table X. *Proportionality between hypoxanthine activity and apparent flavinphosphate content at different stages of purity*

(Corran, Dewan, Gordon & Green [1939]. *Biochem. J.* **33**, 1694.)

Stage	1	2	2/1
	Hypoxanthine activity units per ml.	Apparent flavinphosphate μg. per ml.	
1st (NH$_4$)$_2$SO$_4$ fractionation	5·2	26·4	5·1
After alcohol precipitation	5·8	27·4	4·7
After 1st alumina adsorption	4·0	16·4	4·1
After 2nd alumina adsorption	3·2	14·2	4·4
Extinction* ratio 8·3	4·3	22	5·1
,, ,, 7·4	8·8	41·6	4·7
,, ,, 6·4	5·0	22	4·4
,, ,, 6·2	3·3	15·3	4·6
		Average	4·6

$$* \text{ Extinction ratio} = \frac{E\ 270\ m\mu.}{E\ 450\ m\mu.}$$

Table X shows the proportionality between the hypoxanthine activity of various preparations and the concentration of bound flavin. There is a strict correlation from the crudest preparations to the highest purity level. The ratio of the three activities, hypoxanthine : aldehyde : dihydrocoenzyme I, is also constant throughout purification. It follows therefore that the three activities are functionally associated with the same flavoprotein.

Table XI summarizes the catalytic constants of milk flavoprotein. The three activities are approximately of the same order of magnitude, aldehyde being the most reactive of the three substrates. The method of measuring catalytic activity needs some consideration at this point. Milk flavoprotein in impure preparations catalyses the oxidation of hypoxanthine by molecular oxygen. With purification direct reaction with oxygen becomes progressively more erratic, and at the highest purity level no aerobic manometric measurements can be made. This effect may be interpreted to mean either that the flavoprotein becomes

Table XI. *Summary of catalytic constants of milk flavoprotein*

(Corran, Dewan, Gordon & Green [1939]. *Biochem. J.* **33**, 1694.)

	Hypoxanthine	Acetaldehyde	Dihydro-coenzyme I
Relative activities (hypoxanthine = 100)	100	186	85
Turnover number	306	570	260
$Q_{M.B.}^{H_2}$ * per mg. flavin-phosphate	9×10^5	$1 \cdot 67 \times 10^6$	$7 \cdot 6 \times 10^5$
$Q_{M.B.}^{H_2}$ per mg. protein	2420	3500	2060

* $Q_{M.B.}^{H_2} = \mu l. H_2$ transferred to methylene blue per hour per mg. dry weight.

non-autoxidizable on purification or that the enzyme is protected against H_2O_2 in the crude but not in the purified state. With either aldehyde or dihydrocoenzyme I as substrate the reaction with oxygen is of a negligible magnitude even in the crude state of the enzyme. The most reliable and consistent method of measuring velocities has been found to be the anaerobic Thunberg method using methylene blue instead of oxygen as the oxidizing agent. The Michaelis constant (Km) and the kinetics of the three substrates are not identical. The above constants apply only to certain specified experimental conditions at which the three reactions proceed optimally. With variation of the experimental conditions the ratio of the three activities does not remain constant.

Liver flavoprotein (aldehyde oxidase)

It has long been recognized that the aerobic oxidation of aldehydes in liver is catalysed by (1) the xanthine-aldehyde oxidase and (2) a specific aldehyde oxidase. In 1939 Gordon, Green & Subrahmanyan(50) isolated from pig liver a flavoprotein which specifically catalysed the oxidation of aldehydes to their corresponding acids. Unlike its counterpart in milk the liver aldehyde oxidase showed no activity towards either hypoxanthine or dihydrocoenzyme I.

The following is a résumé of the method of preparation. Minced pig liver is diluted with water, and alcohol added to make the final concentration 20 %. The suspension is heated to 48° for 6 min. and the denatured protein precipitate is centrifuged off and discarded. Basic lead acetate is added to the clear supernatant fluid; the lead precipitate is centrifuged off and decomposed with $M/2$ Na_2HPO_4. The lead-free supernatant fluid is made 40 % saturated with respect to $(NH_4)_2SO_4$. The precipitate is then dissolved in water. To attain higher purity levels some five fractionations in ammoniacal $(NH_4)_2SO_4$ between the limits of 20–35 % saturation are carried out. No procedure has yet been found for eliminating traces of catalase present as impurity. The strong light absorption of residual catalase at 410 mμ. has frustrated attempts to obtain the visible absorption spectrum of the flavoprotein.

The prosthetic group of liver flavoprotein is flavinadenine dinucleotide, though the flavin moiety accounts for only $\frac{1}{3}$rd the total light absorption at 450 mμ. The chemical nature of the other coloured group or groups is unknown.

On addition of acetaldehyde to a strong solution of the flavoprotein, part of the brown colour is "instantaneously" bleached. Vigorous shaking with air restores the original colour. This cycle of reduction and oxidation can be repeated many times. Spectrophotometric measurement shows that the partial bleaching of the colour corresponds in magnitude to the absorption due to the flavin group. In fact the characteristic band of flavin at 450 mμ. disappears on reduction and reappears on shaking with air. At 38° and pH 7·2 each molecule of the enzyme is reduced and oxidized ca. 550 times per min.

The enzyme catalyses the oxidation of aldehydes directly by molecular oxygen. Other hydrogen acceptors can be substituted for molecular oxygen, e.g. methylene blue, cytochrome c, nitrate, etc.

The enzyme is fragile. Desiccation or brief exposure to dilute

104 FLAVOPROTEIN ENZYMES

cyanide (M/100) irreversibly eliminates catalytic activity without denaturing in a gross way the flavoprotein. The enzyme is stable in dilute ammoniacal $(NH_4)_2SO_4$; but in water or in slightly acid or neutral salt solutions, the catalytic activity rapidly falls off even at 0°.

REFERENCES

(1) Warburg & Christian [1932]. *Biochem. Z.* 254, 438.
(2) Kuhn, Györgyi & Wagner-Jauregg [1933]. *Ber. chem. Ges.* 66, 317; 66, 576; 66, 1034.
(3) Kuhn, Reinemund, Kaltschmitt, Ströbele & Trischmann [1935]. *Naturwissenschaften*, 23, 260.
Kuhn, Reinemund, Weygand & Ströbele [1935]. *Ber. chem. Ges.* 68, 1765.
(4) Karrer, Schöpp & Benz [1935]. *Helv. chim. Acta*, 18, 426.
(5) Michaelis, Schubert & Smythe [1936]. *Science*, 84, 2171; *J. biol. Chem.* 116, 587.
(6) Michaelis & Schubert [1938]. *Chem. Rev.* 22, 437.
Michaelis [1935]. *Chem. Rev.* 16, 243.
(7) Warburg & Christian [1933]. *Biochem. Z.* 257, 492.
(8) Kuhn, Rudy & Weygand [1936]. *Ber. chem. Ges.* 69, 1543.
(9) Straub [1938]. *Nature, Lond.*, 141, 603.
(10) Warburg & Christian [1938]. *Biochem. Z.* 295, 261; 298, 150.
(11) Theorell [1935]. *Biochem. Z.* 275, 344; 278, 263.
(12) Warburg & Christian [1938]. *Biochem. Z.* 298, 368.
(13) Kuhn & Rudy [1936]. *Ber. chem. Ges.* 69, 2557.
(14) Corran & Green [1938]. *Biochem. J.* 32, 2231.
(15) Ball [1938]. *J. Biol. Chem.* 128, 51.
(16) Kuhn & Desnuelle [1937]. *Ber. chem. Ges.* 70, 1907.
(17) Warburg [1938]. *Ergebn. Enzymforsch.* 7, 210.
(18) Haas [1938]. *Biochem. Z.* 298, 378.
(19) Krebs [1935]. *Biochem. J.* 29, 1620.
(20) Keilin & Hartree [1936]. *Proc. Roy. Soc.* B, 119, 114.
(21) Negelein & Brömel [1939]. *Biochem. Z.* 300, 225.
(22) Corran, Green & Straub [1939]. *Biochem. J.* 33, 793.
(23) Dewan & Green [1938]. *Biochem. J.* 32, 626.
(24) Euler & Hellström [1938]. *Hoppe-Seyl. Z.* 252, 31.
Adler, Euler & Hellström [1937]. *Sr. Vet. Akad. Ark. Kemi*, 12, 1.
(25) Straub [1939]. *Biochem. J.* 33, 787.
(26) Keilin & Smith [1939]. *Nature, Lond.*, 143, 333.
(27) Fischer & Eysenbach [1937]. *Ann. Phys., Lpz.* 30, 99.
(28) Fischer, Roedig & Rauch [1939]. *Naturwissenschaften*, 27, 196.
(29) Morgan, Stewart & Hopkins [1922]. *Proc. Roy. Soc.* B, 94, 109.
(30) Schardinger [1902]. *Z. Untersuch. Nahr.- u. Genussm.* 5, 1113.
(31) Booth [1935]. *Biochem. J.* 29, 1732.
(32) Dixon & Thurlow [1924]. *Biochem. J.* 18, 976.
(33) Coombs [1927]. *Biochem. J.* 21, 1259.
(34) Booth [1938]. *Biochem. J.* 32, 503.
(35) Booth [1938]. *Biochem. J.* 32, 494.

(36) Dixon [1925]. *Biochem. J.* **19**, 507.
(37) Harrison [1929]. *Biochem. J.* **23**, 982.
     Harrison & Thurlow [1926]. *Biochem. J.* **20**, 217.
(38) Green [1934]. *Biochem. J.* **28**, 1550.
(39) Filitti [1935]. *J. chim. Phys.* **32**, 1.
(40) Corran, Dewan, Gordon & Green [1939]. *Biochem. J.* **33**, 1694.
(41) Philpot [1938]. *Biochem. J.* **32**, 2240.
(42) Dixon [1938]. *Enzymologia*, **5**, 198.
(43) Dixon & Keilin [1936]. *Proc. Roy. Soc.* B, **119**, 159.
(44) Adler, Euler & Günther [1939]. *Nature, Lond.*, **143**, 641.
(45) Bernheim & Bernheim [1934]. *J. biol. Chem.* **107**, 275.
(46) Krebs [1939]. *Enzymologia*, **7**, 53.
(47) Haas, Horecker & Hogness [1939]. *J. biol. Chem.* **130**, 425.
(48) Dixon & Kodama [1926]. *Biochem. J.* **20**, 1104.
(49) Philpot [1939]. *Biochem. J.* **33**, 1707.
(50) Gordon, Green & Subrahmanyan [1940]. *Biochem. J.* **34**, 764.

CHAPTER V

Cu protein enzymes, with an addendum on a Zn protein enzyme

Haemocyanin, the respiratory pigment of certain arthropods and molluscs, is a compound of copper with a specific protein. It can form in air a loose compound with molecular oxygen which dissociates at lower oxygen tensions. The oxygenated compound (oxyhaemocyanin) is blue and the non-oxygenated compound is colourless. Haemocyanin may be considered as the prototype of the family of Cu proteins in the same sense that haemoglobin is the prototype of the Fe porphyrin proteins. Copper is found in nature as the prosthetic group of haemocyanin, polyphenol oxidase, laccase, monophenol oxidase (tyrosinase), haemocuprein and hepatocuprein. The function of haemocyanin is that of oxygen transport. Polyphenol oxidase, monophenol oxidase and laccase are enzymes which catalyse the oxidation of certain benzenoid compounds. The functions of haemocuprein and hepatocuprein have yet to be elucidated. There is evidence that the enzyme in plants which catalyses the oxidation of ascorbic acid is also a Cu protein[18, 19].

Strictly speaking, there is not one haemocyanin but a group of chemically distinct haemocyanins. The percentage of copper in the haemocyanins of different species[1] varies from 0·17 to 0·38 (cf. Table XII). Haemocyanins also differ in solubility, crystallizability, sensitivity to acid, etc.

Table XII. *Copper content of different haemocyanins*

(Data collected by Redfield, Coolidge & Shotts [1928]. *J. biol. Chem.* **76**, 194, and by Kubowitz [1938]. *Biochem. Z.* **299**, 32.)

Source of haemocyanin	% Cu
Octopus	0·25 (0·38)*
Helix pomatia	0·25 (0·29)*
Loligo pealei	0·26 (0·34)*
Limulus polyphemus	0·17
Homarus americanus	0·19 (0·34)*
Dromia vulgaris	0·17

* Data in parenthesis taken from Redfield *et al.*

Copper is present in haemocyanin in the cupro form. The formation of a compound with molecular oxygen does not involve oxidation of the copper. In other words, oxyhaemocyanin is an oxygenation and not an oxidation product of haemocyanin. The oxidation of cuprous haemocyanin to the cupric form [2] (methaemocyanin) can be carried out with powerful oxidizing agents. Unlike methaemoglobin, methaemocyanin can combine reversibly with molecular oxygen.

CO combines reversibly with haemocyanin in the ratio of one molecule of CO per two atoms of copper [3,13]. This complex is not dissociated by light but is dissociated at low partial pressures of CO. Cyanide reacts irreversibly with haemocyanin by splitting off the copper (Kubowitz [3]). Dialysis of a solution of haemocyanin, which has been treated with cyanide, yields the copper-free protein no longer capable of combining with molecular oxygen. Addition of copper in the form of a cuprous salt regenerates the original functional haemocyanin [3]. This experiment demonstrates that copper is the prosthetic group of haemocyanin. Acid treatment similarly liberates copper from haemocyanin, though splitting by this means is irreversible.

There is no rule about the colour of Cu proteins:

	Cupro	Cupric
Oxyhaemocyanin	Blue	—
Haemocyanin	Colourless	—
Polyphenol oxidase	Colourless	Pale yellow
Hepatocuprein	Colourless	Colourless
Haemocuprein	Colourless	Deep blue
Ascorbic oxidase	—	Green

Those Cu proteins with pronounced colour show no specific absorption bands in the visible region of the spectrum, absorption being diffuse.

Polyphenol oxidase and laccase undergo a cycle of transformation from the cupric to the cupro form, and vice versa, in the course of their catalytic activity. These valency changes occur with extraordinary rapidity.

The chemical and physical properties of the polyphenol oxidase vary from species to species. For example, the copper content of the potato oxidase [3] is 0·2 %, whereas that of the cultivated mushroom, *Agaricus campestris* [4], is 0·3 %. Apart from differences in the copper content, the relative speeds with which the various benzenoid compounds are attacked vary with the source of the polyphenol oxidase. It is quite in order to speak of a family of polyphenol oxidases sharing certain general catalytic properties but differing in details. Polyphenol oxidase and laccase are plant enzymes. Similar enzymes (7) have been described in the tissues of arthropods and certain molluscs, and in the ink sac of cephalopods. There is no evidence of the enzymes in vertebrate tissues. Haemocuprein and hepatocuprein are the only general representatives which are known in vertebrate tissues of Cu proteins, although their catalytic function, if any, is still unknown.

Polyphenol oxidase (potato)

To Kubowitz (3) falls the distinction for the first experimental proof that copper is to be included among the prosthetic groups of oxidation enzymes. In the course of purifying the polyphenol oxidase of potato he found a linear relation between the copper

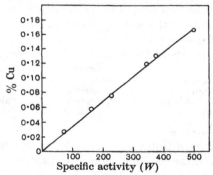

Fig. 14. The proportionality between the copper content and the activity of preparations of the potato polyphenol oxidase. (Kubowitz [1937]. *Biochem. Z.* **292**, 221.) Specific activity $W = \mu l.$ O_2 per mg. enzyme per min.

content of his preparations and the catalytic activity of the enzyme (cf. Fig. 14). He obtained a preparation of the enzyme with a copper content of 0·2 % and clearly demonstrated that the metal was the prosthetic group of the enzyme.

The following is a résumé of Kubowitz's (3) method of preparation. The press juice of minced potato peels is mixed with acetone (final vol. 37 %) and the crude enzyme is precipitated. Then follows precipitation of the enzyme from 68 % saturated $(NH_4)_2SO_4$ solution. The precipitate is redissolved in water and the enzyme again thrown out with 37 % acetone. The enzyme is adsorbed on and eluted from $Al(OH)_3$. For final purification fractional acetone precipitation between the limits of 27–44 % is carried out. The yield of enzyme by this method is *ca.* 0·5 %.

The potato polyphenol oxidase catalyses the oxidation by molecular oxygen of *o*-diphenols such as catechol, adrenaline, dihydroxyphenylalanine[5], etc. The oxidation of catechol is the fastest. Monophenols such as phenol and *p*-cresol are oxidized, but not immediately. A lag period from a few minutes to several hours elapses before oxidation of the monophenol begins [5, 6]. We shall consider elsewhere the significance of the lag period in the oxidation of the monophenols. Resorcin, hydroquinone and ascorbic acid are not oxidized by the enzyme.

The catalytic oxidation of catechol by molecular oxygen takes place in two distinct stages:

(1) [catechol with OH, OH groups] + 2 cupric enzyme ⟶ [o-quinone with =O, =O groups] + 2 cupro enzyme

(2) 4 cupro enzyme + O_2 → 4 cupric enzyme

The enzyme undergoes a cycle of reduction by catechol and oxidation by molecular oxygen. No H_2O_2 is produced in the reaction.

Kubowitz[3] employed the following system for testing the activity of the potato oxidase: hexosemonophosphate and its specific protein enzyme, coenzyme II, catechol, potato oxidase and molecular oxygen. The essential reactions are:

(1) hexosemonophosphate + coenzyme II

$\xrightarrow[\text{enzyme}]{\text{hexosemonophosphoric}}$ phosphohexonate + reduced coenzyme II

(2) reduced coenzyme II + *o*-quinone → coenzyme II + catechol

(3) catechol + O_2 $\xrightarrow{\text{potato enzyme}}$ *o*-quinone

The rate of oxidation of hexosemonophosphate is determined by the rates with which *o*-quinone is reduced by dihydrocoenzyme II and with which catechol is oxidized by molecular oxygen. The overall reaction therefore is a measure of the activity of the polyphenol oxidase. The reader may wonder why so complicated a system is necessary to test the activity of the potato enzyme. Kubowitz maintains that the enzyme shows highest activity when its substrate, catechol, is present in catalytic amounts. The function of the hexosemonophosphoric system is merely to regenerate catechol from its corresponding quinone. In that way the activity of the enzyme in presence of minute amounts of catechol can be measured.

One mg. of the purified enzyme catalyses the uptake of $575\,\mu l$. O_2 per min. $(Q_{O_2} = 34,500)$. The rates of reduction of the cupric enzyme by catechol and oxidation of the cupro enzyme by oxygen may be represented by the following equations:

$$-\frac{d\,(\text{cupric})}{dt} = k_{\text{red.}}\,[\text{cupric}]\,[\text{catechol}]$$

$$-\frac{d\,(\text{cupro})}{dt} = k_{\text{ox.}}\,[\text{cupro}]\,[O_2]$$

where brackets indicate concentrations of the reactants. At $20°$,

$$k_{\text{red.}} = 4 \times 1{\cdot}48 \times 10^7 \times \frac{1}{\dfrac{\text{mols catechol} \times \text{min.}}{\text{litre}}}$$

and $\qquad k_{\text{ox.}} = 4 \times 3{\cdot}56 \times 10^7 \times \dfrac{1}{\dfrac{\text{mols } O_2 \times \text{min.}}{\text{litre}}}.$

The velocity constants for other substrates, particularly monophenols, are much smaller.

For the reversible resolution the enzyme is reduced with catechol and treated with excess cyanide anaerobically. The enzyme is then precipitated by 83 % saturation with $(NH_4)_2SO_4$, centrifuged and washed with saturated $(NH_4)_2SO_4$ containing $0{\cdot}2\ M$ KCN. After dialysing the solution at least 93 % of the copper is removed. Neither the copper-free protein nor cupric salts are able to catalyse the oxidation of catechol. On addition of an equivalent amount of copper in the form of $CuSO_4$ to the protein solution 50 % of the original enzyme activity is restored. Some ten times the theoretical amount of $CuSO_4$ is necessary to restore 98 % of the original activity. Kubowitz interprets this discrepancy to mean that only part of the copper combines with the protein at the "catalytic point of attachment", the rest is bound by groups not concerned in enzymic activity. Fe, Co, Ni, Mn and Zn are inactive as prosthetic groups.

Under anaerobic conditions the system of catechol and its enzyme absorbs CO (Keilin(15)):

catechol + 2 cupric enzyme → quinone + 2 cupro enzyme

2 cupro enzyme + CO → CO—(cupro enzyme)

The absorption of CO was adduced by Kubowitz as proof that the enzyme is being reduced, since only cupro compounds combine with CO. On further addition of cyanide the CO compound of the enzyme is decomposed and the original amount of CO gas is evolved. In absence of catechol the enzyme does not combine with CO. Hence the enzyme as isolated must be in the cupric state.

Polyphenol oxidase (Agaricus campestris)

Keilin & Mann(4) isolated from cultivated mushrooms a highly purified polyphenol oxidase containing 0·3 % copper. Table XIII contains a summary of the essential stages and of the degree of purification attained at each stage. From 15 kg. of mushrooms the yield of homogeneous enzyme is *ca.* 10 mg. The activity of the purest preparation is some 6000 times greater than that of mushroom and 2100 times greater than that of the first crude extract. The Q_{O_2} of the best preparations is *ca.* $1 \cdot 1 \times 10^6$. 1γ Cu of this enzyme in 1 min. at 20° transfers to catechol *ca.* 6000 μl. O_2. The Q_{O_2} of Cu would be *ca.* $3 \cdot 6 \times 10^9$, which is some twenty times the figure obtained by Kubowitz for the potato enzyme. Fig. 15 shows that the proportionality between the copper content and activity is apparent only after considerable purification. This means that the bulk of the copper initially present in the mushroom has no relation to the enzyme.

The crude extracts of the mushroom catalyse the oxidation of monophenols such as *p*-cresol and tyrosine. With successive purification the power of oxidizing monophenols gradually diminishes and in the purest preparations it is practically nil. Furthermore, the oxidation of monophenols is characterized by

Table XIII. *Stages in the purification of catechol oxidase*
(Keilin & Mann [1938]. *Proc. Roy. Soc.* B, 125, 187.)

Stage	En-zyme units	Q_{O_2}	Purpuro-gallin number
A. Crude expressed juice of mush-room (15 l.)	320	785	0·45
B. After $(NH_4)_2SO_4$ precipitation (4 l.)	320	6,600	3·3
C. After lead acetate treatment (7·8 l.)	170	—	14
D. After absorption on and elution from $Ca_3(PO_4)_2$ gel. (30 ml.)	75	462,000	310
E. After fractional precipitation with a mixture of lead acetate and alcohol (30 ml.)	36	560,000	380
F. After further fractional precipitation with lead acetate (30 ml.)	34	870,000	550
G. After adsorption on and elution from $Ca_3(PO_4)_2$ gel. (21 ml.)	21	—	670
H. After fractional precipitation with lead acetate and acetone (6 ml.)	9	1,170,000	940

Purpurogallin number is mg. purpurogallin formed from pyrogallol in 5 min. at 20° per 1 mg. of dry weight of enzyme preparation.

1 enzyme unit (E.U.) is the amount which produces 1000 mg. of purpurogallin in 5 min. at 20° C.

$Q_{O_2} = \mu l.\, O_2$ absorbed per mg. dry wt. of enzyme per hour in presence of catechol at 37°.

a lag period which may be a matter of minutes or hours, depending upon the purity of the enzyme preparation. These facts may indicate either that the oxidation of monophenols is catalysed by the enzyme working in collaboration with some additional factor which is gradually lost in the process of purification, or that there is an enzyme distinct from the catechol oxidase which specifically attacks monophenols. The hypothesis that the catechol oxidase plus some factor is equivalent to the monophenol oxidase is not altogether satisfactory in absence of any direct evidence of the existence of such a factor, or of its nature. Richter[5], in an analysis of the nature of the lag period in the oxidation of monophenols by the potato enzyme, found that substances which combine with *ortho*quinones strongly inhibited monophenol oxidation and extended the lag period, whereas substances which gave rise to *ortho*quinones such as catechol and adrenaline diminished or abolished the lag period. These results

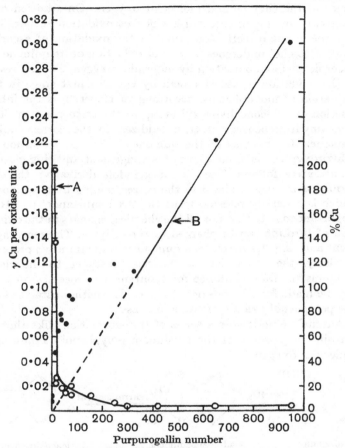

Fig. 15. The proportionality between the copper content and the activity (P.N.) of the mushroom polyphenol oxidase. Curve A (o——o) refers to the γ Cu per oxidase unit; curve B (•——•) refers to the % Cu. (Keilin & Mann [1938]. *Proc. Roy. Soc.* B, **125**, 187.)

led him to the conclusion that *ortho*quinones are responsible for oxidation of monophenols in the same way as they are necessary for oxidation of ascorbic acid by the oxidase system.

The polyphenol oxidase offers many possibilities of successive oxidations. For example, ascorbic acid is not oxidized in presence of the enzyme and molecular oxygen. On addition of a catalytic amount of catechol a vigorous oxidation ensues. The enzyme is not directly concerned in the oxidation of ascorbic acid. Catechol undergoes a cycle of reduction of its quinone by ascorbic acid and oxidation by molecular oxygen. The net result is the oxidation of ascorbic acid by O_2. The first stage in the oxidation of monophenols, according to Raper [7], is the introduction of a second hydroxyl group in the *ortho* position. The *ortho*dihydroxybenzene is then oxidized to the corresponding quinone. In the case of the quinone formed by oxidation of dihydroxyphenylalanine (dopa) rearrangement and ring closure immediately follows. The dihydroxyindole derivative thereby formed is in turn oxidized to the corresponding indolequinone, which is generally referred to as the "red substance" [7]. It is important to note that not all the chemical changes in the oxidation of tyrosine can be charged to the enzyme. The rearrangement and ring closure of dopa quinone is a spontaneous process; similarly, the conversion of the red substance to the black melanins [7]. Due allowance for spontaneous processes must in fact be made for all oxidations of phenols whether catalysed by the polyphenol oxidase or by peroxidase.

Adrenaline undergoes a series of transformations like that of tyrosine in presence of the mushroom polyphenol oxidase and molecular oxygen [8].

adrenaline   adrenalinequinone   leuco-adrenochrome

leuco-adrenochrome   adrenochrome

Adrenochrome can also be formed by oxidation of adrenaline with the cytochrome oxidase—cytochrome c—oxygen system (8). Adrenaline is oxidized by cytochrome c to adrenochrome and in turn ferro-cytochrome c becomes oxidized by molecular oxygen in presence of the cytochrome oxidase. In reconstructed systems from animal tissues adrenochrome has been shown by Green & Richter (8) to act as a "carrier" in the oxidation of metabolites by molecular oxygen via coenzyme I.

Laccase

The latex of an Indo-Chinese lacquer tree (*Rhus succedanea*) contains an enzyme which catalyses the oxidation of different polyphenols. To Bertrand (9) we owe the first extensive study of this enzyme and of related enzymes in other plants. He showed that the darkening and hardening of latex is due to the catalytic oxidation of urushiol and laccol, polyphenols present in the latex.

Keilin & Mann (10) in 1939 isolated a blue Cu-protein from the latex of *Rhus succedanea* which shows all the properties of the enzyme in the crude latex. The acetone powder of the latex was extracted with water and the enzyme was purified by a method similar to that used by Keilin & Mann (4) in the purification of the polyphenol oxidase of the cultivated mushroom.

The best preparations of the enzyme contain 0·154 % Cu. There is evidence that 55 % of the dry weight is accounted for by a polysaccharide impurity. This is in agreement with the fact that the N content is only 6·45 %. Assuming that the polysaccharide is an impurity not connected with the enzyme the Cu content of the homogeneous laccase protein would be about 0·34 %.

Strong solutions of laccase are distinctly blue in colour (10). On addition of a small amount of the substrate, e.g. ascorbic acid or hydroquinone, the blue colour is bleached. Shaking the leuco-enzyme with air restores the initial blue colour. The colour irreversibly disappears on boiling the enzyme or treating it with cyanide and diethyl-dithio-carbamate. These reagents split off the copper from combination with the protein.

Laccase differs in several respects from the polyphenol oxidase of cultivated mushrooms (10). Laccase oxidizes p-phenylenediamine better than catechol, whereas polyphenol oxidase scarcely oxidizes p-phenylenediamine and shows maximum activity with catechol. Purified laccase does not oxidize p-cresol at all, whereas polyphenol oxidase shows a very slight tendency to oxidize monophenols after a protracted lag period. Crude laccase preparations oxidize ascorbic acid directly, although with purification this oxidation becomes progressively slower. Polyphenol oxidase, whether crude or pure, oxidizes ascorbic acid only through a carrier like catechol. Finally, laccase is not poisoned by CO, whereas polyphenol oxidase is strongly inhibited.

With p-phenylenediamine as substrate at 20° the Q_{O_2} of pure laccase is *ca.* 40,000. The turnover number of the enzyme is therefore *ca.* 2500 per min. The activity of laccase is about the same as that of the polyphenol oxidase of potato, but only $\frac{1}{20}$th the activity of the polyphenol oxidase from cultivated mushrooms.

The blue colour of laccase obtained from *Rhus succedanea* was at first ascribed to the copper prosthetic group by analogy with haemocuprein. Further investigation by Keilin & Mann (17) has shown that the blue pigment which invariably accompanies laccase can be obtained free of copper and protein though always associated with carbohydrate. The pigment preparation separated from the enzyme can be reversibly reduced to its leuco compound and irreversibly decolorized by strong acid and by boiling. As yet, the authors have been unable to prepare laccase free of this blue pigment, and they suggest the possibility that the pigment is in some way connected with the normal activity of the enzyme and may function either as an additional prosthetic group or as a carrier between the enzyme and certain substrates.

Laccase has also been prepared from the latex of the Japanese and Burmese lacquer trees (*Rhus vernicifera* and *Melanorrhoea usitata*). These laccases are also Cu-proteins associated with a blue pigment. The Cu content of the best preparations is *ca.* 0·24 %.

Monophenol oxidase (Lactarius piperatus)

Dalton & Nelson (16) have isolated from the press juice of wild mushrooms (*Lactarius piperatus*) a copper protein which catalyses the oxidation by molecular oxygen of monophenols to their corresponding o-quinones. Their best preparation contains 0·23 % bound Cu. One mg. of the enzyme at this stage catalyses an oxygen uptake of 600,000 μl. per hr.

The enzyme can be induced to crystallize when highly purified. But crystallization is attended by great loss in catalytic activity and profound alteration in solubility. After several weeks the crystals become insoluble even in $N/10$ soda.

The *Lactarius* enzyme oxidizes catechol at $\frac{1}{10}$th the rate of p-cresol. The ratio $\dfrac{p\text{-cresol activity}}{\text{catechol activity}}$ remains constant throughout the purification procedure. The comparative inactivity of the *Lactarius* enzyme towards polyphenols and the comparative inactivity of the *Agaricus* and potato enzymes towards monophenols suggests that the alterations in ratio

$$\frac{\text{monophenol activity}}{\text{polyphenol activity}},$$

observed by workers in the course of isolating phenol oxidases, are referable to separation of two distinct enzymes rather than to loss of hypothetical factors.

Haemocuprein

Mann & Keilin [11] have isolated from the red blood corpuscles and serum of ox, sheep and horse a crystalline blue Cu protein which accounts for the total copper present in blood. The best preparations contain 0·34 % Cu and appear homogeneous in an ultra-centrifugal field (mol. wt. *ca.* 35,000). In acid solution the Cu is irreversibly split off from the denatured protein. Fresh preparations of haemocuprein do not contain free inorganic Cu. Only in solutions kept for a long time or damaged in any way does the Cu begin to appear in a free state.

The blue colour of haemocuprein is much stronger than that of an equivalent amount of a copper salt. The colour disappears when the protein is denatured or when treated with $Na_2S_2O_4$, which reduces the copper to the cuprous form. The reduction with $Na_2S_2O_4$ is irreversible. The cupro form of haemocuprein cannot be oxidized either by molecular oxygen or by H_2O_2.

There is no information yet available as to the physiological function of this Cu protein. The justification for including this compound in a chapter devoted to Cu protein enzymes is two-fold. In the first place, feeding experiments with higher organisms[12] have revealed the importance of copper in the formation of haemoglobin, cytochrome, cytochrome oxidase, etc. There is obviously some link between copper and the Fe porphyrin catalyst. Since haemocuprein and a related compound in liver account for a very substantial part of the total copper in the organism it is not far-fetched to expect some important role of haemocuprein in the metabolism of certain oxidation enzymes. The second justification is that haemocuprein may be the proto-type in animal tissues of a family of Cu proteins, of which some might show activity as oxidation catalysts.

Hepatocuprein

Mann & Keilin[11] have also isolated a Cu protein compound from liver which is not identical with haemocuprein. The liver compound contains 0·34 % Cu and is almost colourless. Unlike haemocuprein it shows no tendency to crystallize. The relation between haemocuprein and hepatocuprein is not clear. They may either be two different proteins or the same protein with the copper in a different valency state.

Ascorbic oxidase (squash)

In Chapter VIII ascorbic oxidase is discussed in full. We shall merely mention here that Lovett-Janison & Nelson[19] have isolated from squash (*C. pepo condensa*) a Cu protein which catalyses the oxidation of ascorbic acid by O_2. The best preparations which are *ca.* 1000 times more active per mg. dry weight than the original material contain 0·15 % Cu. At high concentrations the enzyme solutions are green, at lower concentrations blue or bluish green.

Addendum on a zinc protein

Keilin & Mann[14] have isolated from red blood corpuscles a Zn protein compound which appears to be identical with carbonic anhydrase, the enzyme which catalyses the reaction

$$H_2CO_3 \rightleftharpoons CO_2 + H_2O.$$

There is a parallelism between the carbonic anhydrase activities and zinc contents at all stages from the starting material to the final preparation. The yield from 10 l. ox blood was 250 mg. of colourless protein containing 14·95 % N and 0·31–0·34 % Zn. No other metal was found to be present in significant quantity· "Zinc belongs to the universally distributed micro-constituents of plants and animals and is therefore much more widely distributed in nature than carbonic anhydrase. Like iron and copper, zinc probably forms with different proteins several compounds having different properties and functions." It is in the realm of possibility that some of the oxidation enzymes might have Zn as a prosthetic group. In any event, the discovery of yet another metal which can function as a prosthetic group of an enzyme should stimulate closer investigation of the form in which many trace metals occur in living tissues.

REFERENCES

(1) Hernler & Phillippi [1933]. *Hoppe-Seyl. Z.* 216, 110.
Roche & Jean [1934]. *Bull. Soc. Chim. biol.* 16, 769.
Redfield, Coolidge & Shotts [1928]. *J. biol. Chem.* 76, 194.
(2) Conant, Chow & Schoenbach [1933]. *J. biol. Chem.* 101, 463.
(3) Kubowitz [1938]. *Biochem. Z.* 299, 32; [1937]. *Biochem. Z.* 292, 221.
(4) Keilin & Mann [1938]. *Proc. Roy. Soc.* B, 125, 187.
(5) Richter [1934]. *Biochem. J.* 28, 901.
(6) Graubard & Nelson [1935]. *J. biol. Chem.* 111, 761.
(7) Raper [1932]. *Ergebn. Enzymforsch.* 1, 270.
(8) Green & Richter [1937]. *Biochem. J.* 31, 596.
(9) Bertrand [1894]. *C.R. Acad. Sci., Paris,* 118, 1215.
(10) Keilin & Mann [1939]. *Nature, Lond.,* 143, 23.
(11) Mann & Keilin [1938]. *Proc. Roy. Soc.* B, 126, 303.
(12) Elvehjem [1931]. *J. biol. Chem.* 90, 111.
Cohen & Elvehjem [1934]. *J. biol. Chem.* 107, 97.
Yoshikawa [1937]. *J. Biochem., Tokyo,* 25, 627.
(13) Root [1934]. *J. biol. Chem.* 104, 239.
(14) Keilin & Mann [1939]. *Nature, Lond.,* 144, 442.
(15) Keilin [1929]. *Proc. Roy. Soc.* B, 104, 206.
(16) Dalton & Nelson [1939]. *J. Amer. chem. Soc.* 61, 2946.
(17) Keilin & Mann [1940]. *Nature, Lond.,* 144, 442.
(18) Ramasarma, Datta & Doctor [1940]. *Enzymologia,* 8, 108.
(19) Lovett-Janison & Nelson [1940]. *J. Amer. chem. Soc.* 62, 1409.

CHAPTER VI

Thiaminoprotein enzymes

Williams and his colleagues[1] in 1936 definitely established the constitution of vitamin B_1 (polyneuritic vitamin) and effected its synthesis. The vitamin is now known as thiamine. It consists of a thiazole and pyrimidine nucleus linked by a carbon atom:

$$\begin{array}{cc}
\text{N}\!=\!\!=\!\text{CNH}_2 & \text{N}\!=\!\!=\!\text{CNH}_2 \\
\mid \qquad \mid & \mid \qquad \mid \\
\text{CH}_3\text{C} \quad \text{CCH}_2\text{NH}_2 + \text{HCSSH} \rightarrow & \text{CH}_3\text{C} \quad \text{CCH}_2\text{NHCSH} \\
\parallel \qquad \parallel & \parallel \qquad \parallel \\
\text{N}\!-\!\!-\!\text{CH} & \text{N}\!-\!\!-\!\text{CH}
\end{array}$$

2-Methyl 5-aminomethyl 6-aminopyrimidine

$$\begin{array}{c}
\text{N}\!=\!\!=\!\text{CNH}_2 \\
\mid \qquad \mid \qquad\qquad \text{CH}_2\text{CH}_2\text{OR} \\
\text{CH}_3\text{C} \quad \text{CCH}_2\text{NHCSH} + \quad \mid \\
\parallel \qquad \parallel \qquad\qquad \text{CH}_2\text{ClCOMe} \\
\text{N}\!-\!\!-\!\text{CH}
\end{array}$$

$$\begin{array}{cc}
& \text{CH}_3 \\
\text{N}\!=\!\!=\!\text{CNH}_2\text{HCl} & \text{C}\!=\!\!=\!\text{CCH}_2\text{CH}_2\text{OH} \\
\mid \qquad \mid \qquad \text{Cl} & \mid \\
\rightarrow \quad \text{CH}_3\text{C} \quad \text{C}\!-\!\text{CH}_2\!-\!\text{N} & \\
\parallel \qquad \parallel & \diagdown \\
\text{N}\!-\!\!-\!\text{CH} & \text{C}\!-\!\!-\!\text{S} \\
& \text{H}
\end{array}$$

Thiamine chloride hydrochloride

The above is the method of Todd & Bergel[2] for synthesizing thiamine chloride hydrochloride. The chlorides, bromides, picrolonates, etc. of thiamine show an interesting dimorphism of physical properties, for which there is as yet no satisfactory explanation.

Thiamine[1] is rapidly destroyed in alkaline solution by the rupture of the carbon link between the pyrimidine and thiazole nuclei. Stability is greatest in acid solution. Thiamine is heat unstable at pH 7 but not at pH 3. The absorption spectrum of thiamine shows two bands in the u.v. with peaks at 235 and 267 mμ. respectively. There is no absorption in the visible part of the spectrum.

Oxidation of thiamine with alkaline potassium ferricyanide yields thiochrome, a pale yellow, blue-fluorescing substance.

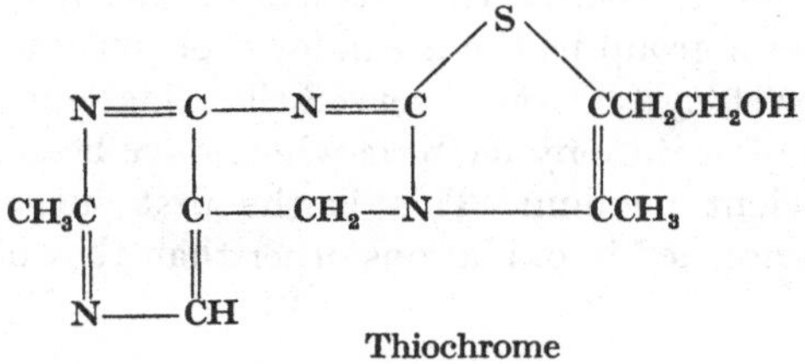
S
N
C
N
C
CCH₂CH₂OH
CH₃C
C
CH₂
N
CCH₃
N
CH

Thiochrome

Thiochrome has been isolated by Kuhn *et al.* (3) from yeast. Synthetic thiochrome prepared by Barger *et al.* (4) has been found identical with the naturally occurring yeast compound.

In alkaline solution thiamine can be reduced with hyposulphite to the corresponding dihydrothiazole form (Lipmann (5)):

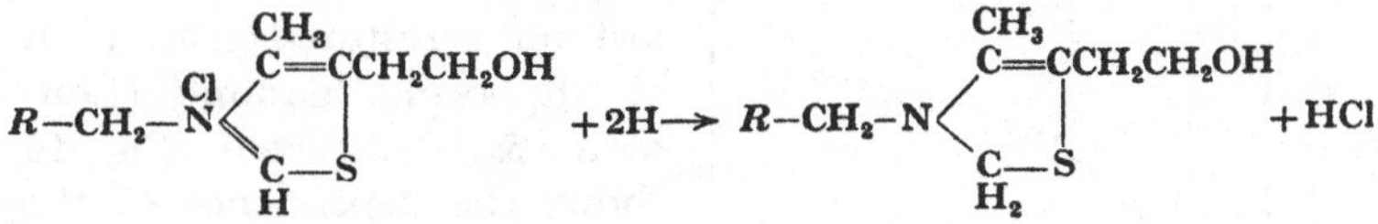
CH₃
C=CCH₂CH₂OH
Cl
R—CH₂—N
C—S
H
+2H→
R—CH₂—N
CH₃
C=CCH₂CH₂OH
C—S
H₂
+HCl

The reduced form is non-autoxidizable in air.

Lohmann & Schuster (6) in 1937 isolated in crystalline form the coenzyme for carboxylase (the yeast enzyme which catalyses the decarboxylation of pyruvic acid). They found the coenzyme to be the pyrophosphoric ester of thiamine. Tauber (7) has synthesized diphosphothiamine by treating thiamine with molten pyrophosphoric acid. Weil-Malherbe (22) has improved the method by treating bromothiamine with acid silver pyrophosphate in pyrophosphoric acid. There is ample evidence of the existence in animal tissues and micro-organisms of enzymes which catalyse the phosphorylation of thiamine as well as the dephosphorylation of diphosphothiamine.

Diphosphothiamine is known to be the prosthetic group of a series of enzymes all of which are concerned in the breakdown of pyruvic acid, e.g. carboxylase of yeast and the pyruvic oxidases of *Bact. Delbruckii*, *Gonococcus* and *Streptococcus haemolyticus*. In animal tissues there is strong evidence that diphosphothiamine is concerned in the oxidation of pyruvic and other α-ketonic acids. As yet no one has succeeded in isolating the protein part of the thiaminoproteins either in micro-organisms or in animal tissues. There is thus little information as to the role of the prosthetic group in the breakdown of pyruvic acid.

Quastel & Webley (23) have observed that thiamine accelerates the oxidation of acetate by bacteria which have been grown in a thiamine-deficient medium. This is the first suggestion that thiamine is concerned in oxidations other than that of α-ketonic acids.

Carboxylase (yeast)

An aqueous suspension of dried yeast rapidly decomposes pyruvic acid with evolution of CO_2. This decarboxylation is catalysed by an enzyme known as carboxylase. If the yeast suspension is thoroughly washed in slightly alkaline solution the power of decomposing pyruvic acid is lost, but is restored on addition of diphosphothiamine (Lohmann & Schuster (6)). Fig. 16 shows the dependence of the rate of CO_2 evolution on the concentration of added diphosphothiamine. Both thiamine and monophosphothiamine are unable to replace the diphospho-compound in this reaction. The reaction falls off rapidly with time. This has been assumed to be due to the inhibitory effect of the product of decarboxylation, viz. acetaldehyde. Both Mg and Mn specifically accelerate the catalytic decarboxylation (cf. Table XIV). Mn is ca. five times as

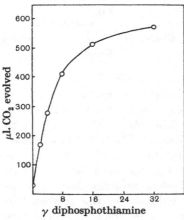

Fig. 16. Carboxylase activity of alkaline-washed dried yeast as a function of the diphosphothiamine concentration. (Lohmann & Schuster [1937]. *Biochem. Z.* **294**, 188.)

efficient as Mg per mg. dry weight. Very likely, however, Mg is concerned physiologically, since it occurs in considerable concentration in yeast, whereas Mn is present only in traces. The

role of Mg or Mn and, for that matter, the role of diphosphothiamine in the catalytic decarboxylation of pyruvic acid in
yeast has yet to be clarified.

Table XIV. *The effect of Mg and Mn on
the activity of carboxylase*

(Lohmann & Schuster [1937]. *Biochem. Z.* **294,** 188.)

Addition		μl. CO_2
3 γ diphosphothiamine		34
,,	+1 γ Mg	38
,,	+30 γ Mg	116
,,	+100 γ Mg	175
,,	+1 γ Mn	50
,,	+10 γ Mn	165

Long & Peters[8] have found that carboxylase of washed yeast
in presence of diphosphothiamine can catalyse the decarboxylation of α-ketobutyric and α-ketovaleric acids as well as pyruvic
acid. Carboxylase can therefore be considered as specific for the
class of α-ketomonocarboxylic acids.

Ochoa[9] observed that thiamine, although inactive by itself,
considerably enhances the catalytic effect of a given amount of
diphosphothiamine on the decarboxylation of pyruvic acid by
alkaline-washed yeast. Long & Peters[8] have confirmed this
strange effect of thiamine using α-ketobutyric and α-ketovaleric
acids respectively as substrates for yeast carboxylase:

	μl. CO_2 production in 30 min.			
	No addition	10 γ thiamine	1 γ diphospho-thiamine	10 γ thiamine +1 γ diphos-phothiamine
Blank	1	6	1	1
α-Ketobutyrate	22	27	165	548
α-Ketovalerate	17	11	122	386
Pyruvate	27	31	181	465

Weil-Malherbe[22] attributes this effect to phosphorylation of
thiamine under the conditions of the experiment. Purified preparations of yeast carboxylase[22] do not respond to the presence
of thiamine either in presence or absence of diphosphothiamine.

Recently Green *et al.*[24] have obtained carboxylase from top yeast in highly purified form. The enzyme is a diphosphothiamine metalloprotein. The best preparations contain 0·46 % diphosphothiamine and 0·13 % magnesium. One mg. of purified enzyme catalyses the formation of 12,140 μl. CO_2 per hour at 30°. By precipitation with ammoniacal $(NH_4)_2SO_4$ the specific protein may be separated from the prosthetic group and metal. To reconstitute the original enzymic activity, (1) the specific protein, (2) diphosphothiamine, (3) some divalent metal are necessary. Diphosphothiamine cannot be replaced by thiamine, monophosphothiamine, coenzyme 1, adenineflavinedinucleotide or adenosinetriphosphate. All divalent metals tested, e.g. Mg, Mn, Fe··, Ca, Cd, Zn, and Co, were active in the system though with varying efficiency. Monovalent and trivalent cations were inactive. The conditions for the quantitative reconstitution of carboxylase from its component parts are complex. In high salt concentrations carboxylase is a firmly bound conjugated protein, whereas in dilute salt solutions or at alkaline pH it dissociates almost completely. It is probable that the metal acts as a "cement" substance binding the protein to the prosthetic group.

Pyruvic oxidase (bacteria)

An aqueous suspension of acetone dried *Bact. Delbruckii* hardly takes up oxygen in presence of pyruvic acid. On addition of a few γ of diphosphothiamine a large increase in the rate of oxygen uptake ensues (Lipmann(10)). Simultaneous addition of flavinadenine dinucleotide increases the rate still more. Lipmann(11) succeeded in bringing the pyruvic enzyme into solution by extracting the "acetone" powder with phosphate buffer and centrifuging off the insoluble residues. Repeated precipitation of the enzyme from solution at 50 % saturation $(NH_4)_2SO_4$, pH 3 and 0° yields a preparation of the enzyme which catalyses the oxidation of pyruvic acid only in presence of both diphosphothiamine and flavinadenine dinucleotide:

Diphosphothiamine	—	15γ	15γ
Flavinadeninedinucleotide	20γ	—	20γ
μl. O_2	15	5	248

Diphosphothiamine cannot be replaced by thiamine, and the dinucleotide cannot be replaced by riboflavin, riboflavinphosphate, or the Warburg-Christian flavoprotein. The oxidation of pyruvic acid can also be conducted anaerobically provided some suitable hydrogen acceptor such as methylene blue is present. It is important to note that pyruvic acid is not being decarboxylated as in the case of yeast. The process is more correctly an oxidative decarboxylation, i.e. aldehyde, the product of decarboxylation, becomes oxidized to acetic acid.

Yeast:
$$CH_3COCOOH \rightarrow CO_2 + CH_3CHO$$

Bact. Delbruckii:
$$CH_3COCOOH \xrightarrow[M.B.]{O_2} CO_2 + CH_3COOH$$

It is significant that flavinadenine dinucleotide should be necessary for the oxidative decarboxylation but not for the straightforward decarboxylation. Very likely it is concerned as prosthetic group of the enzyme which catalyses the oxidation of acetaldehyde (or some derivative thereof) to acetic acid. Provisionally we may consider diphosphothiamine as the prosthetic group for the decarboxylating enzyme. The complication, however, is that if there are a decarboxylating and oxidative enzyme respectively, there is as yet no evidence that the process of decarboxylation can proceed independently of the subsequent oxidation.

The rate of dehydrogenation of pyruvate was found by Lipmann[20] to be proportional to the concentration of phosphate within certain limits. This specific effect of phosphate was explained in terms of the following scheme:

pyruvate + phosphate + 2H → acetylphosphate + CO_2
acetylphosphate + H_2O → acetate + phosphate

According to this view the phosphorylation of pyruvate takes place simultaneously with the oxidative decarboxylation. The product of oxidation acetylphosphate can easily be transformed in presence of appropriate enzymes into acetate and phosphate.

Barron[12,13] has obtained a preparation of the pyruvic enzyme by allowing suspensions of Gonococcus to autolyse in slightly alkaline solution. The enzyme catalyses the aerobic oxidation of pyruvic acid to acetic acid and CO_2. There is evidence that diphosphothiamine is the prosthetic group of this enzyme. The sensitivity of the *Gonococcus* pyruvic enzyme to various reagents is unlike that of yeast carboxylase. Low concentrations of α-naphthol, HCN and NaF (0.01–0.001 M) inhibit almost completely the bacterial enzyme although without any effect on the yeast enzyme.

Krebs [15] and Barron & Lyman [13] have studied the anaerobic dismutation of pyruvic acid in *Staphylococcus* and *Gonococcus*:

$$2CH_3COCOOH \rightarrow CH_3COOH + CO_2 + CH_3CHOHCOOH$$

One molecule of pyruvic acid is oxidized to acetic acid and CO_2, whilst another molecule is reduced to lactic acid. We may think of the dismutation of pyruvic acid as a carrier-linked reaction between the pyruvic enzyme system on the one hand and the lactic enzyme system on the other. Presumably, some carrier is reduced by pyruvate (catalysed by the pyruvic enzyme) and the reduced carrier is in turn oxidized by another molecule of pyruvate (catalysed by the lactic enzyme working in reverse). The nature of the carrier is obscure. Lipmann [14] suggests that some flavin-adenine protein compound is acting as carrier in the dismutation. In bacteria as well as in animal tissues either the direct oxidation of pyruvic acid by oxygen or the anaerobic dismutation can take place depending upon conditions such as the oxygen tension, presence of carriers, etc.

Pyruvic oxidase (animal tissues)

Peters and his colleagues [16] have observed in the course of some years of study that the minced brain of avitaminous pigeons oxidizes pyruvic acid subnormally. This power is restored by addition to the mince of thiamine in very minute concentration. This discovery was in fact made long before the constitution of thiamine was established and before cocarboxylase was shown to be a phosphoric derivative of thiamine. Some progress is now being made in the way of extracting the enzyme responsible for the oxidation of pyruvic acid in animal tissues. Recently Banga *et al.* [17, 19] have succeeded in extracting the pyruvic enzyme from avitaminous brain. The preparation requires the following additions to enable the oxidation of pyruvic acid to take place: diphosphothiamine, fumarate, adenylic acid, Mg or Mn (21), and phosphate. Adenylic acid, phosphate and fumarate seem to be concerned not in the primary oxidation of pyruvic acid (presumably to acetic acid), but in the secondary oxidation of whatever is the actual oxidation product of pyruvic acid in this system. Although thiamine is active in the mince it shows little or no activity in the isolated system. Clearly thiamine must undergo rapid phosphorylation under physiological conditions.

Krebs & Johnson (18) have described enzymes present in tissue slices which catalyse either the aerobic oxidation of pyruvic acid to acetic acid and CO_2 or the anaerobic dismutation of pyruvic acid to CO_2, acetic and lactic acids. This finding. in conjunction with those of Peters, indicate a system in animal tissues closely related to that of the pyruvic oxidase in bacteria.

REFERENCES

(1) Williams & Spies [1938]. *Vitamin B_1*, New York.
    Williams & Cline [1936]. *J. Amer. chem. Soc.* 58, 1504.
    Cline, Williams & Finkelstein [1937]. *J. Amer. chem. Soc.* 59, 1052.
(2) Todd & Bergel [1937]. *J. chem. Soc.* p. 364.
(3) Kuhn, Wagner-Jauregg, van Klaveren & Vetter [1935]. *Hoppe-Seyl. Z.* 234, 196.
(4) Barger, Bergel & Todd [1935]. *Nature, Lond.,* 136, 259; *Ber. chem. Ges.* 68, 2257.
(5) Lipmann [1936]. *Nature, Lond.,* 138, 1097.
(6) Lohmann & Schuster [1937]. *Biochem. Z.* 294, 188.
(7) Tauber [1938]. *J. Amer. chem. Soc.* 60, 730.
(8) Long & Peters [1939]. *Biochem. J.* 33, 759.
(9) Ochoa [1938]. *Nature, Lond.,* 141, 831.
(10) Lipmann [1937]. *Enzymologia,* 4, 65.
(11) Lipmann [1939]. *Nature, Lond.,* 143, 436.
(12) Barron [1936]. *J. biol. Chem.* 113, 695.
    Barron & Miller [1932]. *J. biol. Chem.* 97, 691.
(13) Barron & Lyman [1939]. *J. biol. Chem.* 127, 143.
(14) Lipmann [1939]. *Nature, Lond.,* 143, 281.
(15) Krebs [1937]. *Biochem. J.* 31, 661.
(16) Peters [1936]. *Biochem. J.* 30, 2206.
    Peters & Thompson [1934]. *Biochem. J.* 28, 916.
(17) Banga, Ochoa & Peters [1939]. *Nature, Lond.,* 143, 764; 144, 74.
(18) Krebs & Johnson [1937]. *Biochem. J.* 31, 645.
(19) Banga, Ochoa & Peters [1939]. *Biochem. J.* 33, 1109; 33, 1980.
(20) Lipmann [1939]. *Nature, Lond.,* 144, 381.
(21) Ochoa [1939]. *Nature, Lond.,* 144, 834.
(22) Weil-Malherbe [1939]. *Biochem. J.* 33, 1997.
(23) Quastel & Webley [1939]. *Nature, Lond.,* 144, 633.
(24) Green, Herbert & Subrahmanyan [1940]. *J. biol. Chem.* (in press).

CHAPTER VII

Cytochrome-reducing dehydrogenases

Knowledge of the chemical nature of the prosthetic group is complete for relatively few oxidation enzymes. At the present stage of the subject it is clearly impossible to classify all oxidation enzymes on a chemical basis. In this chapter we shall consider a group of oxidation enzymes which show striking similarities from a functional point of view. These enzymes catalyse the oxidation of their respective substrates by various hydrogen acceptors, such as methylene blue and other oxidation-reduction indicators but not by molecular oxygen. We shall apply the term dehydrogenase to the members of this group, since they are independent of any external coenzyme for activity. Thus far, none of them have been resolved into a protein and prosthetic group. Crude preparations of these dehydrogenases catalyse with extraordinary efficiency the oxidation of their respective substrate by the three cytochrome components. This property is in fact diagnostic of the group. The mechanism of the reaction of these systems with the cytochromes has yet to be clarified. On purification or manipulation two enzymes of this group are known to lose the capacity to react with cytochrome c, although still capable of reacting with methylene blue. There is a suggestion that cytochrome b is the intermediary link between the succinic enzyme system and cytochrome c. That is to say, the oxidation of succinic acid by cytochrome c takes place in at least two stages,

$$\text{succinic} \rightarrow \text{cytochrome } b \rightarrow \text{cytochrome } c$$

the arrows indicating the direction of transfer of hydrogen. Thus the group of cytochrome-reducing dehydrogenases should more correctly be referred to as cytochrome b-reducing dehydrogenases. The whole question is still *sub judice* and the final decision of nomenclature must await further progress.

α-Glycerophosphoric dehydrogenase

Animal tissues contain two enzymes which catalyse the oxidation of α-glycerophosphate. One is a pyridinoprotein enzyme which has been described in Chapter III. The other is independent of the pyridine nucleotides and conforms in properties to the general characteristics of the cytochrome-reducing group of enzymes.

The α-glycerophosphoric dehydrogenase has been systematically studied by Green[1]. It is found in practically every tissue of the animal body in relatively high concentration. Skeletal muscle is one of the richest sources of the enzyme, whereas cardiac muscle contains a comparatively small amount.

The enzyme is prepared from rabbit skeletal muscle as follows. The muscles are minced and washed thoroughly with tap water to remove all soluble material. The mince is then ground thoroughly with sand and $M/50$ phosphate buffer pH 7·2. After filtering off the insoluble debris through muslin the filtrate is acidified to pH 4·6 with acetic acid and the precipitate is centrifuged off. The precipitate which contains the active enzyme is resuspended in phosphate buffer pH 7·2.

The enzyme specifically catalyses the oxidation of $l(+)$-α-glycerophosphate ($CH_2OHCHOHCH_2OPO_3H_2$) to glyceraldehydephosphate ($CHOCHOHCH_2OPO_3H_2$). $d(+)$-α-Glycerophosphate, glycerol, β-glycerophosphate, 2-phosphoglycerate and 3-phosphoglycerate are not attacked.

A mixture of the dehydrogenase as prepared above and α-glycerophosphate hardly reacts with oxygen in absence of a carrier. The addition of a dilute solution of cytochrome c accelerates the reaction enormously (cf. Fig. 17). The rate of oxidation is proportional to the concentration of cytochrome only within the limit of the saturating concentration. Calculation has been made of the number of times that cytochrome c is reduced and oxidized per minute in the α-glycerophosphate system. Fig. 18 shows the dependence of the turnover number on the concentration of cytochrome c. The maximum value attained is about 300.

In addition to cytochrome c a large list of hydrogen acceptors

have been found to be active, e.g. methylene blue, benzyl vio-
logen and other oxidation-reduction indicators. There is, however, a fundamental distinction between the action of cytochrome
and that of other hydrogen acceptors. In the former case the
collaboration of the dehydrogenase and the cytochrome oxidase

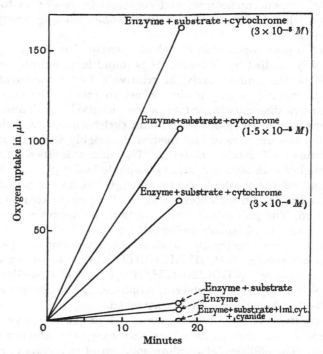

Fig. 17. Effect of cytochrome c on the enzymic oxidation of α-glycerophosphate. (Green [1936]. *Biochem. J.* **30**, 629.)

is involved; in the latter case only the dehydrogenase is involved.
This can easily be demonstrated by the effect of cyanide. With
methylene blue as the carrier, cyanide even in high concentrations ($M/10$) has no inhibitory effect; whereas with cytochrome
c as the carrier, $M/1000$ cyanide completely inhibits all oxidation. Thus it is clear that the inhibition by cyanide is connected
with some enzyme other than the dehydrogenase. There is strong

evidence that the cytochrome oxidase is in fact the cyanide-sensitive enzyme. The oxidase invariably accompanies the α-glycerophosphoric enzyme as prepared above from animal

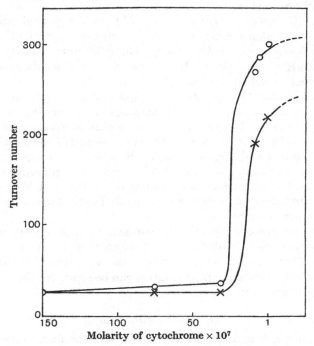

Fig. 18. The turnover number of cytochrome c in the α-glycerophosphate enzyme system as a function of concentration. o———o yeast. ×———× heart. (Green [1936]. *Biochem. J.* **30**, 629.)

tissues. The reaction with molecular oxygen via the cytochromes may therefore be represented as follows:

$$\alpha\text{-glycerophosphate} + 2 \text{ cytochrome} \xrightarrow{\text{dehydrogenase}} \text{glyceraldehyde-}$$
$$\text{phosphate} + 2 \text{ reduced cytochrome}$$

$$4 \text{ reduced cytochrome} + O_2 \xrightarrow[\text{oxidase}]{\text{cytochrome}} 4 \text{ cytochrome}$$

Flavin, yeast flavoprotein, heart flavoprotein, glutathione, ascorbic acid and adrenaline cannot act as carriers even when present in high concentration. Pyocyanide is the only other naturally occurring carrier (*Bact. pyocyaneus*), besides cytochrome *c*, which has been shown to be active in the α-glycerophosphoric system.

The Michaelis constant Km is 0·01 M for *dl*-α-glycerophosphate. There is no sharp maximum *p*H. Within the range of *p*H 6–11, the velocity varies only slightly. The maximum velocity is at about *p*H 8; for a given *p*H the velocity also depends upon the chemical nature of the buffer used.

The product of oxidation, viz. glyceraldehydephosphate, is highly unstable and rapidly decomposes into methylglyoxal and phosphoric acid. The presence of considerable amounts of methylglyoxal is therefore always observed at the end of the oxidation. The isolation and identification of the unstable triosephosphate has not yet been accomplished. The evidence is in favour of the view that aldotriosephosphate is actually formed. However, the possibility that the ketotriosephosphate is also formed has not been entirely excluded.

The α-glycerophosphoric dehydrogenase is attached to insoluble particles. Centrifugation of the enzyme suspension brings down all the active material. The process of peptizing the enzyme and then precipitating from solution can be repeated many times without serious loss in activity.

Succinic dehydrogenase

Thunberg[2] and Batelli & Stern[3] were the first to describe the enzyme in animal tissues which catalyses the oxidation of succinic acid ($COOHCH_2CH_2COOH$). Einbeck[4] later showed that fumaric acid ($COOHCH=CHCOOH$) was the product of oxidation. The succinic enzyme is found in practically every tissue of the animal body. Its distribution is also very general among bacteria, moulds and higher plants.

An active preparation of the succinic dehydrogenase can be made from the heart muscle[5] of pig, ox or sheep by the same method used for the preparation of the α-glycerophosphoric enzyme. The succinic enzyme prepared by this method is accompanied by the cytochrome oxidase as well as by cytochromes *a* and *b*. This is to be expected, since all these components are associated with insoluble particles and hence do not yield to any appreciable purification. It is difficult to decide whether the insoluble particles represent the enzyme proper or a bit of the cellular fabric to which the enzyme is attached. It may be more than a coincidence that the two cytochrome-reducing dehydrogenases of animal tissues as well as the cytochrome oxidase and the cytochrome components a and b are all insoluble.

The succinic dehydrogenase in presence of succinate reduces considerable concentrations of oxidized cytochrome c almost instantaneously (Keilin (6)). In fact there is little to choose between hyposulphite and the succinic system as reducing agents for cytochrome c. The cytochromes a and b present in enzyme preparations are also rapidly reduced on addition of succinate. The tests must be made anaerobically to prevent reoxidation of the cytochromes by molecular oxygen—a process catalysed by the cytochrome oxidase.

The suspension of heart muscle prepared as above contains the three cytochrome components (mostly a and b) and the cytochrome oxidase in addition to the succinic enzyme. On addition of succinate to the enzyme preparation reaction with molecular oxygen takes place through the intermediation of the cytochromes, which undergo a cycle of reduction by succinate (catalysed by the dehydrogenase) and oxidation by molecular oxygen (catalysed by the oxidase). Reagents which poison the oxidase, such as cyanide, azide and sulphide, stop the reaction with molecular oxygen by preventing the oxidation of the reduced cytochromes and thereby breaking the catalytic cycle. The reaction can be restored by addition of some substance, such as methylene blue, which can be reduced by succinate in presence of the enzyme and which in the reduced form is autoxidizable in air. The difference between the reaction of succinate with molecular oxygen via the cytochromes or via methylene blue may be summarized in the following way:

Cytochrome

$$\text{succinate} + 2 \text{ cytochrome} \xrightarrow{\text{dehydrogenase}} \text{fumarate}$$
$$+ \text{ reduced cytochrome}$$
$$2 \text{ reduced cytochrome} + \tfrac{1}{2}O_2 \xrightarrow{\text{oxidase}} 2 \text{ cytochrome}$$

Methylene blue

$$\text{succinate} + \text{methylene blue} \xrightarrow{\text{dehydrogenase}} \text{fumarate}$$
$$+ \text{methylene white}$$

$$\text{methylene white} + O_2 \rightarrow \text{methylene blue} + H_2O_2$$

Preparations of the succinic enzyme which already contain the cytochrome components show an increase in the velocity of reaction with molecular oxygen on addition of cytochrome c. That indicates that the amount of cytochrome present is not sufficient to allow the reaction to proceed at maximal velocity. It is possible to make preparations of the succinic enzyme, particularly

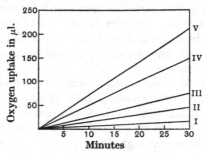

Fig. 19. Effect of concentration of cytochrome c on the rate of oxygen uptake in the succinic enzyme system prepared from liver. I, control without cytochrome; II, 1.9×10^{-5} M cytochrome c; III, 8.8×10^{-5} M; IV, 1.5×10^{-4} M; V, 3×10^{-4} M. (Ogston & Green [1935]. *Biochem. J.* **29**, 1989.)

from liver (5) or kidney (26) which are practically free from the cytochromes. In these cases no appreciable reaction with oxygen takes place unless cytochrome c is added to the system (cf. Fig. 19).

Keilin & Hartree (26) have found that successive freezing and thawing of a succinic preparation of heart muscle eliminates the direct reaction of succinate with molecular oxygen, although the cytochromes present are apparently unaffected by this treatment. Addition of cytochrome c restores the reaction with molecular oxygen.

Under maximal conditions each molecule of cytochrome c is reduced and oxidized 1420 times per min. For the determination of the turnover number Keilin & Hartree[26] added very small quantities of cytochrome c ($ca.$ $2 \cdot 15 \times 10^{-9}$ g. atom Fe) to excess of the succinic enzyme and determined the increase in the oxygen uptake during the first 10 min. of the reaction.

Experiments of Hopkins, Lutwak-Mann & Morgan[7] and Keilin & Hartree[26] have thrown some light on the reaction of succinate with cytochrome c. Preparations of the enzyme acidified to pH 5 for 1 hr. at room temperature lose the power to react with oxygen via the cytochromes although able to react via methylene blue. The inability to work through the cytochrome path cannot be explained in terms of the destruction of the cytochrome oxidase by acid treatment, since the oxidase suffers practically no loss in activity. After acid treatment the reduction either of added cytochrome c or of cytochrome c normally present is no longer catalysed by the enzyme preparation in presence of succinate. Similar effects are obtained by treating the enzyme preparation with pancreatin or alcohol. These experiments indicate that the reduction of cytochrome c by the succinic enzyme system is not direct but proceeds through some intermediary link which is destroyed by exposure to acid, alcohol or pancreatin. The fact that in all the preparations which have lost the property of reducing cytochrome c the component b seems to be invariably affected has led to the suggestion that cytochrome b probably forms a link between the enzyme and cytochrome c.

No other dicarboxylic acid, with the exception of methyl succinic acid, is attacked by the succinic enzyme (Thunberg[8]). Malonic, maleic and oxaloacetic acids although inactive as substrates show the phenomenon of competitive inhibition (Quastel[9]). Some investigators have considered inhibition by $M/100$ malonic acid as a diagnostic test for the succinic enzyme.

The dehydrogenase can catalyse not only the oxidation of succinate to fumarate but also the reverse reduction. The direction in which the reaction will proceed is determined by oxidation-reduction equilibria. An equimolecular mixture of succinate and cytochrome c will react in presence of the enzyme until the former is practically completely oxidized and the latter reduced. There is no tendency for the reaction to reverse. Methylene blue and succinate react only partially, an equilibrium being reached at which definite amounts of all the components are present. That this equilibrium is real can be demonstrated by starting with leuco-methylene blue and fumarate, and obtaining the same equilibrium concentrations of reactants. Thunberg (10), Quastel (11), Lehmann (12) and Borsook (13) have all contributed to the study of the reversibility of the succinic enzyme.

Hopkins *et al.* (24, 25) have carried out experiments which have an important bearing on the nature of the active groups of the succinic dehydrogenase. Incubation of the enzyme with high concentrations of oxidized glutathione (GSSG) renders it completely inactive. Activity can be restored by subsequent treatment with reduced glutathione (GSH). Alloxan can also inactivate the enzyme and here again the effect is reversed by treatment with GSH. Dibasic acids are known from competitive inhibition experiments to combine with the enzyme (e.g. fumaric, malonic and pyrophosphoric (20) acids). Although these acids are not attacked by the enzyme they are able to protect it against inactivation by GSSG. Malonic acid is the most efficient protective agent, being effective in a concentration as low as 0·0002 M. These observations are interpreted by Hopkins *et al.* to mean that some sulphydryl group is intimately connected with the activity of the succinic enzyme. When this group is oxidized by agents such as GSSG or alloxan the activity of the enzyme is lost, but can be restored by reducing the oxidized group to the original —SH form. The fact that the substrate of the enzyme, as well as structurally related acids, protect the —SH group against oxidation is interpreted to mean that the —SH group is normally concerned in the specific combination between enzyme and substrate.

Lactic dehydrogenase (yeast)

The yeast lactic enzyme catalyses the oxidation of lactic and α-hydroxybutyric acids to the corresponding keto acids (14). Only the l stereochemical forms are attacked. Malic, β-hydroxybutyric and γ-hydroxybutyric acids are inactive as substrates.

Crude preparations of the lactic enzyme (5) catalyse the oxidation of lactate by molecular oxygen in presence of added cytochrome *c*. The reaction is cyanide sensitive, indicating that cytochrome c is undergoing a cycle of reduction by the lactic enzyme system and oxidation by the cytochrome oxidase-oxygen system.

Dixon & Zerfas(15) have reported briefly on the properties of a preparation of highly purified yeast lactic enzyme. Unlike the succinic and α-glycerophosphoric enzymes of animal tissues, the yeast enzyme is soluble in water or salt solutions and at no stage is associated with insoluble particles.

No external coenzyme is necessary for its activity. Whereas crude preparations of the yeast lactic dehydrogenase catalyse the oxidation of lactate by cytochrome c as well as by methylene blue, the purified enzyme cannot utilize cytochrome c effectively as hydrogen acceptor even in presence of a cytochrome oxidase preparation of heart containing cytochrome components a and b. This evidence clearly indicates the existence of a factor present in the crude preparations but either lost or destroyed in the process of purification.

Bernheim(16) has shown that $M/100$ pyruvate inhibits the oxidation of lactate by the yeast enzyme. This inhibition is not due to any equilibrium, but rather to competition between oxidant and reductant for the active group of the enzyme.

Lactic dehydrogenase (Gonococcus)

Barron & Hastings(17) found that a suspension of *Gonococcus* in 0·9 % NaCl retained the power of oxidizing lactate but not glucose or pyruvate when kept several days at 3°. After some weeks the cells cytolyse, but the lactic enzyme activity remains unimpaired. An enzyme preparation can be made by suspending the bacteria in a slightly alkaline medium (*ca.* pH 8). Cytolysis soon ensues. The insoluble debris is centrifuged off, the supernatant fluid containing the active enzyme.

The *Gonococcus* enzyme catalyses the oxidation of lactate, α-hydroxybutyrate and glycerate to the corresponding ketonic acids. Lactate is the most readily attacked substrate of the three. A mixture of the enzyme and lactate reacts directly with molecular oxygen. This direct reaction is inhibited by cyanide, H_2S and valeronitrile. In presence of these inhibitors, no oxidation of lactate occurs unless some suitable carrier such as cresyl blue is provided.

Barron & Hastings explain these results along the following lines. The enzyme solution contains two components: (1) the lactate dehydrogenase and (2) an oxygen-activating mechanism very much akin to the cytochrome-cytochrome oxidase system of animal tissues and yeast. There is thus a complete cytochrome system available for the oxidation of lactate. H_2S, cyanide and valeronitrile inhibit the cytochrome oxidase. Hence in presence of these inhibitors some autoxidizable carriers such as cresyl blue or nicotine-haemin must be provided in lieu of the natural cytochrome mechanism for the reaction with molecular oxygen. Barron & Hastings do not come to any decision as to the identity of their oxygen activation system with the cytochrome-cytochrome oxidase system. But the available data, such as cyanide inhibition, destruction by exposure to 52° for some minutes, insolubility, etc., point to the identity of the Barron & Hastings system with the Keilin cytochrome-oxidase system.

Formic dehydrogenase (Bact. coli)

Stickland (18) was the first to prepare a cell-free extract of *Bact. coli* which could catalyse the oxidation of formate to carbon dioxide:

$$HCOOH + acceptor \rightarrow CO_2 + reduced\ acceptor$$

More recently Gale (19) has studied the formic enzyme in some detail. The method of preparation is as follows. A thick suspension of the bacteria is ground in the Booth-Green crushing mill for 2 hr. The suspension is centrifuged and the supernatant fluid is discarded. The sediment contains the particles with which the formic enzyme is associated. The sediment is suspended in phosphate buffer and incubated with trypsin for *ca.* 24 hr. This treatment cleans up the preparation and yields a transparent suspension of fine particles. There is no loss in activity.

The formic dehydrogenase does not require any external coenzyme. Purified preparations do not catalyse the oxidation of formate by molecular oxygen except in the presence of an added carrier such as methylene blue. Preparations of the enzyme at the stage before tryptic digestion catalyse the oxidation of formate by molecular oxygen in absence of added carrier. This reaction is completely inhibited by $M/1000$ HCN. The crude preparations of the enzyme contain both cytochrome oxidase and cytochrome b in sufficient concentration to account for the reaction with oxygen in absence of added carrier. In the course of trypic digestion the oxidase-cytochrome b system is destroyed. When methylene blue is used as hydrogen acceptor $M/1000$ HCN inhibits to the extent of $ca.$ 8 %. The evidence is in favour of the view that the direct and cyanide-sensitive reaction with oxygen observed in crude preparations of the enzyme involves the cytochrome oxidase-cytochrome b system.

The purified formic enzyme is inactivated by oxygen in presence of its substrate. The higher the oxygen tension the greater the rate of inactivation. Activity can be restored by incubating the inactivated enzyme anaerobically either in presence or in absence of formate. In presence of methylene blue the rate of aerobic inactivation is considerably diminished.

The formic dehydrogenase of *Bact. coli* is always found in association with insoluble particles. Attempts to separate the enzyme from the particles have been uniformly unsuccessful.

Choline dehydrogenase (liver)

Mann & Quastel[21] and Bernheim & Bernheim[22] have discovered in liver brei an enzyme system which catalyses the aerobic oxidation of choline to the corresponding aldehyde:

$$CH_2OHCH_2N(CH_3)_3OH + \tfrac{1}{2}O_2 \rightarrow CHOCH_2N(CH_3)_3OH$$

Mann *et al.*[23] consider that the following components are involved in the aerobic cyanide-sensitive oxidation: choline dehydrogenase, the cytochromes and cytochrome oxidase. Thus far no success has been attained in resolving the system into the component parts. Since the available evidence rules out the possibility of a coenzyme, we may consider provisionally that the choline dehydrogenase satisfies the criteria of the group of cytochrome-reducing dehydrogenases. Arsenocholine is attacked by the same enzyme though somewhat more slowly than choline in equivalent concentration.

REFERENCES

(1) Green [1936]. *Biochem. J.* 30, 629.
(2) Thunberg [1909]. *Skand. Arch. Physiol.* 22, 430; [1910]. *Ibid.* 24, 23; [1911]. *Ibid.* 25, 37.
(3) Batelli & Stern [1910]. *Biochem. Z.* 30, 172.
(4) Einbeck [1919]. *Biochem. Z.* 95, 296.
(5) Ogston & Green [1935]. *Biochem. J.* 29, 1983.
(6) Keilin [1929]. *Proc. Roy. Soc.* B, 104, 206.
(7) Hopkins, Lutwak-Mann & Morgan [1939]. *Nature, Lond.*, 143, 556.
(8) Thunberg [1933]. *Biochem. Z.* 258, 48.
(9) Quastel [1926]. *Biochem. J.* 20, 166.
(10) Thunberg [1925]. *Skand. Arch. Physiol.* 46, 339.
(11) Quastel & Whetham [1924]. *Biochem. J.* 18, 519.
(12) Lehmann [1929]. *Skand. Arch. Physiol.* 58, 173.
(13) Borsook & Schott [1931]. *J. biol. Chem.* 92, 535.
(14) Green & Brosteaux [1936]. *Biochem. J.* 30, 1489.
(15) Dixon & Zervas [1939]. *Nature, Lond.*, 143, 551.
(16) Bernheim [1928]. *Biochem. J.* 22, 1178.
(17) Barron & Hastings [1933]. *J. biol. Chem.* 100, 155.
(18) Stickland [1929]. *Biochem. J.* 23, 1187.
(19) Gale [1939]. *Biochem. J.* 33, 1012.
(20) Leloir & Dixon [1937]. *Enzymologia*, 2, 81.
(21) Mann & Quastel [1937]. *Biochem. J.* 31, 689.
(22) Bernheim & Bernheim [1933]. *Amer. J. Physiol.* 104, 438.
(23) Mann, Woodward & Quastel [1938]. *Biochem. J.* 32, 1024.
(24) Hopkins & Morgan [1938]. *Biochem. J.* 32, 611.
(25) Hopkins, Morgan & Lutwak-Mann [1938]. *Biochem. J.* 32, 1829.
(26) Keilin & Hartree [1940]. *Proc. Roy. Soc.* B. In press.

CHAPTER VIII

Unclassified oxidation enzymes

This chapter deals with some of the unclassified enzymes whose properties are sufficiently well studied to merit consideration. Many of these enzymes are now the subjects of intensive experimentation. No attempt will be made therefore to improvise an *ad hoc* system of classification.

α-Hydroxyglutaric dehydrogenase (heart)

We owe our knowledge of this enzyme to the work of Weil-Malherbe [1]. Pig heart is the most suitable source of active extracts. The tissue is minced, washed and ground with sand. The paste is extracted for 20 min. at 37° with $M/10$ NaHCO$_3$. The mixture is filtered and the filtrate is brought to pH 4·6 with acetic acid. The precipitate is centrifuged and washed with water. It is finally suspended in $M/15$ veronal buffer pH 8·2.

A mixture of enzyme and substrate takes up oxygen if pyocyanine* is present as carrier. No coenzyme is necessary for this reaction. In absence of pyocyanine, a slow oxidation ensues. No carriers other than pyocyanine and related phenazine derivatives have been found to be efficient in this oxidation.

The enzyme catalyses the oxidation of $l(-)$-α-hydroxy-glutarate (COOHCH$_2$CH$_2$CHOHCOOH) to α-ketoglutarate (COOHCH$_2$CH$_2$COCOOH). The $d(+)$-isomer is not attacked. The pH maximum of the reaction lies between 8·0 and 8·5. The half-speed concentration of substrate (Km) is about $M/300$; full saturation of the enzyme is reached with a concentration of $M/30$.

Using pyocyanine as the carrier, Weil-Malherbe [1] observed that in presence of cyanide the theoretical end value of the O$_2$ uptake was exceeded by as much as 50 %. Analysis showed that this was due to H$_2$O$_2$ formation. Reduced pyocyanine on autoxidation yields H$_2$O$_2$. In absence of cyanide H$_2$O$_2$ is decomposed by the catalase present in the enzyme preparation. Cyanide, by

---

* Pyocyanine is a phenazine oxidation-reduction indicator which occurs in certain strains of *Bact. pyocyaneus*. It is reducible by many dehydrogenase systems. The reduced or leuco form is autoxidizable in air.

inhibiting catalase activity, allows an accumulation of H_2O_2, thus accounting for the apparent extra oxygen uptake.

If α-ketoglutarate, β-hydroxybutyrate and coenzyme I are incubated anaerobically with the heart enzyme preparation, oxidation of β-hydroxybutyrate to acetoacetate takes place. In absence of either coenzyme I or α-ketoglutarate this oxidation does not occur. It is clear therefore that acetoacetate formation depends upon the simultaneous reduction of α-ketoglutarate. Coenzyme I is the linking agent between the two enzyme systems involved. This is an additional illustration that certain enzyme systems which do not depend upon the pyridine nucleotides may nevertheless react with them.

The α-hydroxyglutaric dehydrogenase is found principally in the brain, diaphragm muscle, cardiac muscle and kidney of those animals tested.

Amine oxidase

This enzyme has been known under many names merely because various investigators were content to study the oxidation of one substrate or restricted group of substrates without considering the question of specificity. Amine oxidase is identical with (1) tyramine oxidase of Hare (2), (2) aliphatic amine oxidase of Pugh & Quastel (3) and (3) adrenaline oxidase of Blaschko *et al.* (4). The amine oxidase specifically catalyses the oxidation of amines, whether aliphatic, aromatic, N substituted, etc. Amino acids are not attacked. Thus it is clear that the amine and amino-acid enzymes are distinct. The amine oxidase catalyses the direct oxidation of its substrate by molecular oxygen; H_2O_2 is formed in the reaction.

We owe our detailed knowledge of the amine oxidase to the thorough analyses of Pugh & Quastel (5), Richter (6) and Kohn (34). They showed that in the catalytic oxidation of primary, secondary and tertiary amines, in all cases an aldehyde and NH_3 or a lower amine were formed:

$$R\overset{+}{C}H_2NH_3 + H_2O + O_2 \rightarrow RCHO + \overset{+}{N}H_4 + H_2O_2$$

$$R\overset{+}{C}H_2NH_2R^1 + H_2O + O_2 \rightarrow RCHO + \overset{+}{N}H_3R^1 + H_2O_2$$

$$R\overset{+}{C}H_2NH(R^1)_2 + H_2O + O_2 \rightarrow RCHO + \overset{+}{N}H_2(R^1)_2$$

Richter postulates the following mechanism of reaction:

$$RCH_2NH_2 + O_2 \rightarrow RCH = NH + H_2O_2$$
$$RCH = NH + H_2O \rightarrow RCHO + NH_3$$

The amine is dehydrogenated to the corresponding imino-derivative, which spontaneously reacts with water to form an aldehyde and NH_3 (or a lower amine).

The second and tertiary amines offer two possibilities of enzymic attack:

(1) $R^1CHNHCHR^2 \rightarrow R^1CHO + NH_2CHR^2$

(2) $R^1CHNHCHR^2 \rightarrow R^2CHO + NH_2CHR^1$

Where R^2 is a methyl, dimethyl or ethyl group and R^1 is a hydrocarbon of more than three carbon atoms or a benzene nucleus, the rule is that the simplest amine is formed. When the enzyme is presented with two groups of similar size the tendency is for the substance to be attacked extremely slowly, if at all.

Table XV contains a list of some of the amines which have been tested with the amine oxidase. Compounds with a methyl or carboxyl group in the α-carbon atom, such as ephedrine, benzedrine or the α-amino acids, are not oxidized. It should be mentioned in passing that ephedrine[4,53] and benzedrine[54] specifically inhibit the activity of the amine oxidase. Substances such as mescaline or hydrastinine with a bulky side chain are either oxidized extremely slowly or not at all. Tyramine, epinine, arterenol, sympatol, adrenaline and isoamylamine are most rapidly attacked. It is interesting that the size of the side chains plays an important role. Methylamine, ethylamine and propylamine are very slowly oxidized, whereas amylamine and isoamylamine are rapidly oxidized. Too large a side chain, as in heptylamine, tends to decrease activity. Histidine, taurine, cadaverine and putrescine, substances in which the hydrocarbon group is interrupted by a second polar group, are not oxidized.

The amine oxidase shows only partial optical specificity. The natural l-adrenaline is oxidized about twice as fast as the d-enantiomorph. Usually enzyme systems obey the "all or none" law with respect to optical specificity. The asymmetric carbon atom of adrenaline is not concerned in the oxidation. Possibly that is why only partial specificity is shown.

Table XV. *The relative rates of oxidation of various*
amines by the amine oxidase

(Blaschko, Richter & Schlossmann [1937]. *Biochem. J.* 31, 2187.)

Substrate	Tissues tested*	Result	Relative rate
Methylamine	l, i	±	<2
Ethylamine	,,	+	<2
Propylamine	,,	+	7
Butylamine	,,	+	54
Amylamine	,,	+	19
Heptylamine	,,	+	5
isoButylamine	,,	+	9
isoAmylamine	l, k,	+	105
Cadaverine	l, i	−	<2
Choline	,,	−	<2
Acetylcholine	,,	−	<2
d-Ornithine	,,	−	<2
Aniline	l	−	<2
Benzylamine	l, i	+	9
β-Phenylethylamine	,,	+	11
dl-Alanine	i	−	<2
Tyramine	l, i	+	100
l-Adrenaline	,,	+	65
d-Adrenaline	,,	+	45
l-p-Sympatol	,,	+	48
d-p-Sympatol	,,	+	−
dl-m-Sympatol	l	+	59
dl-Arterenol	l, i	+	51
Epinine	,,	+	125
Adrenalone	l	+	30
β-Phenyl-β-hydroxy-ethylamine	,,	+	46
Hydrastinine	,,	±	−
Mescaline	,,	+	5
Tryptamine	l, i	+	87
Hordenine	,,	+	12
Homorenon	l	+	16
dl-Alkamine	l, i	+	12
Dihydroxyphenyl-ethylamine	,,	+	140
ω-Aminoacetocatechol	l	+	22

*l =liver, i =intestine, k=kidney

The crude enzyme can be easily prepared from guinea-pig liver by the following procedure: the tissue is ground with sand and mixed with an equal volume of $M/15$ phosphate buffer pH 7·3. The mixture is centrifuged and the solution dialysed for 5 hr. The enzyme solution has a low blank and is stable at 0° for several days. The enzyme occurs principally in liver and intestine. Kidney, heart, lung, spleen and brain contain smaller amounts of the enzyme[49]. Amine oxidase is found in mammals, birds, reptiles, amphibians and fishes. In the invertebrates active preparations are obtained from echinoderms and molluscs, in contrast to the negative results with representatives of arthropoda, annelida and coelenterata.

Blaschko et al. [4] determined the velocity ratios for sympatol, adrenaline, tyramine, isoamylamine, heptylamine and phenylethylamine in a large series of tissues and animals. Though the ratios on the whole were reasonably constant, in a few cases distinct and reproducible differences were observed. The relative rate of oxidation of isoamylamine, for example, was always faster in the liver than in the intestine. However, when considering the possible differences between enzyme preparations obtained from various tissues, such as concentration of inhibitors, salts, protein and interfering enzymes, it is indeed not surprising that the velocity ratios are not strictly constant. We may assume therefore that the amine oxidase is one enzyme, although its chemical constitution may vary somewhat from one animal to another and possibly from one tissue to another in a manner similar to the haemoglobins.

The amine oxidase is insensitive to 10^{-3} M cyanide. Substances containing a hydrocarbon chain, such as the higher aliphatic alcohols or thymol, are strongly inhibitory.

Little is yet known about the reaction of the amine oxidase with hydrogen acceptors other than oxygen. Philpot[7] has found that with tyramine as substrate, oxidation-reduction indicators could be used as acceptors providing the potential of these dyes was more positive than -0.046 v. Some dyes like methylene blue and toluylene blue were unsuitable because of toxic action.

Blaschko et al. [4] consider the physiological importance of the amine oxidase to be twofold. (1) Amines are produced by the action of bacteria on amino acids in the gut. The function of the amine oxidase in the gut is therefore one of detoxication, i.e. converting toxic amines by oxidation to the relatively non-toxic aldehydes, which can then undergo further change. (2) The amine oxidase may play a significant role in the removal of large doses of adrenaline such as are of pharmacological importance. A fact militating against this assumption is that the concentration of adrenaline at which the amine oxidase works with appreciable velocity is very much in excess of the adrenaline concentration normally present in tissues. The half-speed concentration of adrenaline is about $M/100$.

Urico-oxidase or uricase

Batelli & Stern[8] in 1909 described the properties of an enzyme in animal tissues which catalyses the oxidation of uric acid by molecular oxygen. The reaction proceeds according to the following equation[9]:

$$\text{uric acid} + 2H_2O + O_2 \rightarrow \text{allantoin} + CO_2 + H_2O_2$$

The first stage in the oxidation may consist in the formation of 4 : 5-glycol uric acid. This rearranges to form oxyacetylenediurein carboxylic acid, which spontaneously gives rise to allantoin by decarboxylation. The detailed chemistry of the oxidation is not clear. But it is certain that urico-oxidase is involved not in the actual production of allantoin but in the oxidation of uric acid to the allantoin precursor.

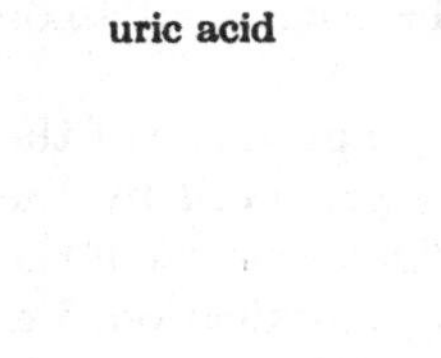

uric acid 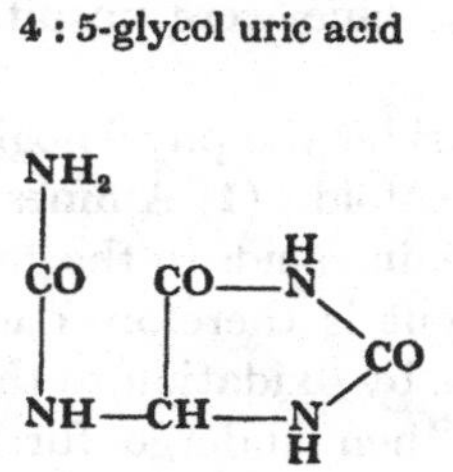   4 : 5-glycol uric acid 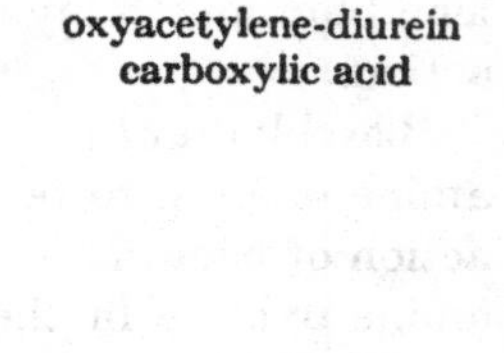   oxyacetylene-diurein
                                                    carboxylic acid

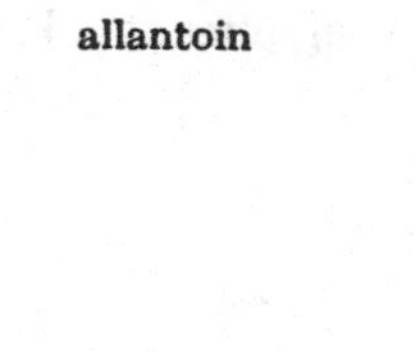

allantoin

Urico-oxidase can be easily and conveniently prepared from fresh ox or pig kidney(9). The tissue is minced in a Latapie, mixed with 6 vol. of acetone and filtered. The acetone treatment is repeated and the residue is dried in an air current. The acetone powder retains activity for months when stored in a vacuum desiccator. Urico-oxidase is associated with insoluble particles which do not dissolve in neutral solution but merely become suspended in the aqueous fluid. Thus on rubbing up the enzyme powder with buffer and then centrifuging hard, all the activity will be found to be in the sediment. The little activity if any present in the supernatant fluid can be removed by high-speed centrifugation.

The enzyme works best in a fairly alkaline medium. The optimum pH is 9·3. Little activity is shown above pH 11 or below 7. No alternative hydrogen acceptor to oxygen has yet been found. There is no information available of the potential level of the system uric acid-allantoin precursor. The negative results obtained with oxidation-reduction indicators may possibly be due to the fact that the potential level of the uric-acid system is too positive relative to those of the indicators.

Uric acid is the only one of a large series of related purine compounds which is attacked by the enzyme(9). The methyl derivatives of uric acid, although not oxidized, specifically inhibit the enzymic oxidation of uric acid. Here again we have evidence that substances structurally related to the substrate may interfere with the activation of the substrate, although they do not undergo activation. The inhibition is due to competition between the substrate and the derivatives for the active group of the enzyme.

Cyanide in a concentration of 0·0003 M inhibits the action of urico-oxidase almost completely. The inhibition is reversible even with much stronger concentrations of cyanide(9). Removal of the added cyanide leads to the restoration of activity. CO, Na_2S, NaF and ethyl urethane have little or no inhibitory action.

H_2O_2 formed in the oxidation of uric acid may be utilized in coupled oxidations(9). If peroxidase and p-phenylenediamine are added to the uric-acid system, the oxygen uptake is exactly twice the theoretical for the oxidation of uric acid to allantoin.

H_2O_2 formed in the reaction oxidizes p-phenylenediamine to coloured oxidation products. This takes place only in presence of peroxidase.

Unlike the xanthine oxidase, the activity of urico-oxidase depends markedly upon the partial pressure of oxygen. The rate of oxidation is approximately twice as fast in pure oxygen as in air. This fact suggests that the active group is only slowly oxidized by molecular oxygen and that the rate of autoxidation varies with the oxygen tension.

Davidson(10) has prepared an extract from liver, 1 mg. dry weight of which absorbs 5000 μl. O_2 per hr. at 38° in presence of uric acid. This activity is some 500 times greater than that of crude acetone powder. The purified preparation was almost colourless. Iron was the only metal found by Davidson in any appreciable concentration (0·15–0·2 %).

Holmberg(48) has purified urico-oxidase a stage farther than that reached by Davidson. The ratio of the iron content to the catalytic activity is much smaller (0·025 %) than that observed by Davidson but not small enough to dismiss the possibility that Fe is connected with the catalytic activity. The presence of 0·13 % Zn is reported in his best preparations. The highly purified enzyme is extremely sensitive to traces of copper, particularly in presence of H_2O_2. It is interesting to note that the ratio $\dfrac{\text{oxygen consumed}}{\text{carbon dioxide produced}}$ for the oxidation of uric acid is identical for crude and highly purified enzyme systems. This clearly indicates that only one enzyme is concerned in the oxidation of uric acid to allantoin and CO_2.

The distribution of urico-oxidase is limited to a few tissues, particularly liver and kidney.

Diamine oxidase

Best & McHenry(12) have described an enzyme in animal tissues which catalyses the oxidation of histamine by molecular oxygen. Edlbacher & Zeller(13) and Zeller(14) have investigated more fully the properties of histamine oxidase and have found that histamine is only one of several diamines which are attacked by the enzyme. They have therefore renamed the enzyme diamine oxidase.

Acetone powder of pig kidney is the starting-point of Zeller's method of preparation. The powder is extracted with weak salt solutions and the extract is dialysed.

The enzyme catalyses the aerobic oxidation of histamine, putrescine [$H_2N(CH_2)_4NH_2$], cadaverine [$H_2N(CH_2)_5NH_2$] and agmatine [$H_2NCH_2NH(CH_2)_4NH_2$] but not of ethylene diamine. Simple amines and amino acids are not attacked. It is thus clear that diamine oxidase is not identical either with the amine or amino-acid oxidase.

The oxidation of histamine was shown by Zeller to conform to the following equation:

$$\begin{array}{c} \text{CH—NH} \\ \text{‖} \quad \text{⟩CH} + O_2 + H_2O \\ \text{C—N} \\ \text{|} \\ \text{CH}_2\text{CH}_2\text{NH}_2 \end{array} \longrightarrow \begin{array}{c} \text{CH—NH} \\ \text{‖} \quad \text{⟩CH} + NH_3 + H_2O_2 \\ \text{C—N} \\ \text{|} \\ \text{CH}_2\text{CHO} \end{array}$$

The production of H_2O_2 was demonstrated by the oxidation of alcohol to aldehyde in presence of catalase and by the production of methaemoglobin from haemoglobin (cf. Bernheim et al.[15]). Zeller et al.[60] have made observations which strongly suggest that the diamine oxidase is a dissociating flavoprotein. The evidence is as yet not conclusive.

Glucose oxidase (Aspergillus)

Müller[16] in 1928 demonstrated the presence of an enzyme in the expressed juice of *Aspergillus niger* which catalyses the oxidation of glucose to gluconic acid by molecular oxygen. The preparation of the enzyme is made as follows. The mould is ground with quartz sand and kieselguhr. The mixture is placed in a hydraulic press and the expressed juice is mixed with several volumes of alcohol. The precipitated enzyme is then dried and powdered.

The glucose oxidase catalyses the oxidation of d-glucose to gluconic acid. d-Mannose and d-galactose are oxidized, though much more slowly than d-glucose. No other monosaccharides or disaccharides are known to be attacked. The enzyme works best at about pH 6·0. The Michaelis constant Km is 0·026 M.

Franke & Lorenz (17) have investigated the kinetics of the *Aspergillus* oxidase. They observed that in presence of H_2S, a reagent which inhibits catalase but without effect on the oxidase, H_2O_2 accumulated almost in quantitative yield. That is to say, for each molecule of glucose oxidized by oxygen one molecule of H_2O_2 was formed. In absence of H_2S only traces of H_2O_2 can accumulate, owing to constant catalytic decomposition of the latter.

Benzoquinone and certain indophenols can replace molecular oxygen as the hydrogen acceptor in the glucose oxidase system. Methylene blue is inactive in this capacity.

HCN, H_2S, azide and hydroxylamine do not inhibit the enzyme. Narcotics have no effect in neutral solution, but inhibit at about pH 4·4. The rate of oxidation is approximately twice as fast in pure oxygen as in air.

There is no evidence as yet that any coenzyme is necessary for this enzyme.

Lactic dehydrogenase (Bact. coli)

Stephenson (57) found that the clarified autolysate of thick *Bact. coli* suspensions contains an active lactic enzyme. In addition to lactate the next homologue α-hydroxybutyrate is oxidized in presence of the enzyme, although much more slowly.

A mixture of the lactic enzyme and lactate fails to absorb oxygen. In presence of oxidation-reduction indicators such as methylene blue an oxygen uptake occurs and lactate becomes oxidized to pyruvate. The properties of the *coli* lactic enzyme closely resemble that of its counterpart in the *Gonococcus*.

Nitratase (Bact. coli) nitrate reductase

Extracts of crushed cells of *Bact. coli* contain an enzyme which catalyses the reduction of nitrate to nitrite (18). If nitrate is added to a mixture of the enzyme preparation and reduced methylene blue (or any other suitable reduced oxidation-reduction indicator) rapid oxidation of the leuco form ensues with simultaneous reduction of nitrate to nitrite:

$$MbH_2 + HNO_3 \rightarrow Mb + HNO_2 + H_2O$$

$$(MbH_2 = \text{reduced methylene blue})$$

No success has as yet attended efforts to effect the converse reaction, i.e. the oxidation of nitrite to nitrate by means of the oxidized forms of oxidation-reduction indicators.

Tarr[19] has discovered an enzyme present in certain strains of the bacterial flora of spoiling fish muscle which catalyses the reduction of trimethylamine oxide $(CH_3)_3NO$ to trimethylamine $(CH_3)_3NH_2$, but not the converse oxidation. The enzyme has not as yet been obtained in cell-free solution. The nature of the physiological reductant for trimethylamine oxide is unknown. Trimethylamine oxide can be replaced by triethylamine oxide though tri-n-propylamine shows only feeble activity.

Dihydroxymaleic oxidase

Szent-Györgyi[20] and his colleagues were led by theoretical considerations to look for an enzyme which catalyses the oxidation of dihydroxymaleic acid. They found such an enzyme to be widely distributed in plant tissues. The dihydroxymaleic oxidase apparently catalyses the oxidation of dihydroxymaleic by molecular oxygen:

$$COOHCOH : COHCOOH + O_2 \rightarrow COOHCOCOCOOH + H_2O_2$$

The product of oxidation, viz. diketosuccinic acid, is unstable and easily decomposes with evolution of CO_2.

Banga & Phillippot[21] later observed that plant peroxidase catalyses the oxidation of dihydroxymaleic acid by H_2O_2. It remained for Theorell & Swedin[22,55] to identify the so-called dihydroxymaleic oxidase with peroxidase. The earlier investigators were misled by the fact that oxygen could be utilized by the enzyme as oxidizing agent in contrast to peroxidase, which is specific for H_2O_2 as oxidizing agent. Theorell & Swedin point out that various reagents such as Mn and HCN stimulate the autoxidation of dihydroxymaleic acid and the H_2O_2 thereby formed can oxidize dihydroxymaleic acid in presence of peroxidase. With dihydroxymaleic acid as substrate peroxidase seems to undergo a cycle of valency change from the ferrous to the ferric condition. In presence of pyrogallol and other phenol substrates peroxidase remains in the ferric state throughout the course of the catalyses. It is premature as yet to decide whether peroxidase can undergo two different catalytic cycles.

Glutamic aminophorase

Braunstein & Kritzmann (23) have discovered in animal tissues and plants a special type of oxido-reduction which involves a simultaneous transfer of an amino group from the reductant to the oxidant. There is apparently a family of enzymes involved in this process of oxido-reductive transamination. In the next chapter we shall consider this process in some detail. As yet only two of the enzymes concerned have been isolated and studied in detail, viz. the glutamic and aspartic aminophorases.

The following is the method of Kritzmann (24) for preparing the glutamic enzyme. Pigeon heart muscle or pig heart is minced and ground with sand. The pulp is extracted with 1% bicarbonate solution. The extract is acidified to pH 4·2 with acetic acid and the precipitate is centrifuged, washed with distilled water and resuspended in phosphate buffer.

The glutamic aminophorase catalyses the following reaction:

$$COOHCHNH_2(CH_2)_2COOH + CH_3COCOOH$$
$$\text{glutamic} \qquad\qquad\qquad \text{pyruvic}$$
$$\rightarrow COOHCO(CH_2)_2COOH + CH_3CHNH_2COOH$$
$$\text{α-ketoglutaric} \qquad\qquad \text{alanine}$$

Glutamate is oxidatively deaminated to α-ketoglutarate, whereas pyruvate is reductively aminated to alanine. It is significant that the reaction is reversible. Fig. 20 shows that the same equilibrium point is reached regardless of whether a mixture of glutamate and pyruvate or of α-ketoglutarate and alanine is employed. Pyruvic acid can be substituted by other monocarboxylic and ketonic acids such as α-hydroxybutyric acid, dimethyl pyruvic acid, etc., but not by dicarboxylic ketonic acids such as oxaloacetic acid or by ketones and aldehydes. Glutamic acid cannot be replaced by any other dicarboxylic amino acid except aminomalonic acid. Only the l enantiomorph of glutamic acid is oxidized. The alanine which is produced by reductive amination of pyruvic acid also belongs to the l stereochemical series. It is interesting to note that minced muscle can transaminate either d or l glutamic acid.

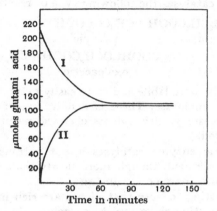

Fig. 20. Time course and reversibility of enzyme action. I, 200 μmoles glutamic and pyruvic acid; II, 200 μmoles α-ketoglutaric and l-(+)-alanine. (Kritzmann [1938]. *Biochimia*, 3, 603.)

Aspartic acid is not directly attacked by the glutamic enzyme. But the transfer of NH_3 from aspartic to pyruvic acid can be catalysed by the enzyme provided a trace of α-ketoglutaric acid is present:

	Trans-amination in μmoles
Enzyme + 200 μmoles glutamic + 200 μmoles pyruvic	85
Enzyme + 200 μmoles aspartic + 200 μmoles pyruvic	0
Enzyme + 200 μmoles aspartic + 1 μmole α-ketoglutaric + 200 μmoles pyruvic	38

The glutamic aminophorase preparation can be purified by exposure to 60° for 20 min. The inactive precipitate which forms is discarded. The enzyme is then precipitated with $(NH_4)_2SO_4$ and the precipitate is dried *in vacuo*. The activity per mg. dry weight of this powder is fifty-six times greater than that of the original muscle tissue. There is evidence that a coenzyme is necessary for the functioning of the glutamic aminophorase.

Aspartic aminophorase

Kritzmann [25] in a preliminary note has announced the isolation from muscle tissue and plants of the aspartic aminophorase. This enzyme catalyses the following type of reaction:

$$COOHCHNH_2CH_2COOH + CH_3COCOOH$$
$$\quad\text{aspartic}\qquad\qquad\qquad\text{pyruvic}$$
$$\rightarrow COOHCOCH_2COOH + CH_3CHNH_2COOH$$
$$\qquad\text{oxaloacetic}\qquad\qquad\qquad\text{alanine}$$

The enzyme is more labile and more easily extractable than the glutamic aminophorase. There is definite evidence that a co-enzyme is necessary, although its chemical nature has not yet been elucidated.

The aspartic enzyme catalyses also the transamination of hydroxyaspartic acid, though more slowly than aspartic acid. Glutamic acid is inactive as substrate.

It is interesting to note that plants are rich in the aspartic aminophorase, though completely lacking in the glutamic enzyme.

Glyoxalase

Glyoxalase is an enzyme which catalyses the conversion of methyl glyoxal to d lactic acid (i.e. the unnatural isomer),

$$CH_3COCHO + H_2O \rightarrow CH_3CHOHCOOH$$

The reaction involves an internal Cannizzaro or intramolecular oxido-reduction; the ketonic group is reduced to the alcohol, while the aldehyde group is simultaneously oxidized to the acid. The enzyme also attacks substituted glyoxals such as phenyl glyoxal.

A highly active enzyme can be prepared by acetone precipitation of the aqueous extract of minced liver. The enzyme is widely distributed in animal tissues and micro-organisms. Its physiological function is not yet understood.

Lohmann [26] discovered that purified preparations of glyoxalase were unable to catalyse the conversion of methyl glyoxal except in presence of the tripeptide, glutathione. Before discussing the role of glutathione in the catalytic reaction we shall first consider the general properties of this somewhat mysterious substance.

Hopkins (27) in 1929, followed by Kendall *et al.* (28), isolated from baker's yeast a tripeptide of glutamic acid, cysteine and glycine—now referred to as glutathione:

$$COOHCHNH_2CH_2CH_2CONHCH(CH_2SH)CONHCH_2COOH$$

The —SH group of glutathione in presence of suitable catalysts, such as metals, etc., can be oxidized by oxygen to the disulphide form:

$$2\ RSH + O_2 \rightarrow RSSR + H_2O_2$$
$$(R = \text{tripeptide residue})$$

The mechanism of this catalysis is complex; the reader is referred to the classical paper of Meldrum & Dixon (29) for a full discussion of the factors concerned in the autoxidation of —SH glutathione.

The distribution of glutathione is very general in animal tissues as well as in plants and bacteria. It is significant that glutathione is present principally if not entirely in the form of the —SH tripeptide. That can mean either that the reduction of oxidized glutathione proceeds much more rapidly than the oxidation of reduced glutathione or that for all practical purposes there is no cycle of reduction and oxidation—glutathione existing only in the reduced form.

Hopkins & Elliott (30) have made a thorough study of the behaviour of glutathione in animal tissues. They found that —SS— glutathione was rapidly reduced by liver mince and that this reduction was enzymic. Even in presence of oxygen glutathione existed almost entirely in the reduced form as long as the reducing systems in liver mince were active. With exhaustion of the substrates of these reducing systems a comparatively slow oxidation of glutathione took place. In other words, reduction of glutathione was a much more rapid process than oxidation, and as long as the reducing systems were functioning glutathione was maintained in the reduced state.

The high concentration and the almost universal distribution of glutathione in cells of all descriptions have led biochemists to an intensive search for its physiological function. The only significant positive finding on the enzymic side as yet is Lohmann's discovery of its role as coenzyme for glyoxalase.

The mechanism of the action of glutathione as coenzyme has been studied by Quastel[31], Giršavičius[32], Yamazoye[33] and others, who have shown that methyl glyoxal reacts with reduced glutathione in presence of the enzyme to form a compound which then dissociates into lactic acid and reduced glutathione. Yamazoye has actually isolated the methyl glyoxal-glutathione compound. It is important to note that this compound is formed

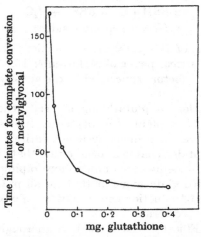

Fig. 21. Dependence of glyoxalase activity on the concentration of reduced glutathione. (Lohmann [1932]. *Biochem. Z.* **254**, 452.)

only in presence of the enzyme. The compound undergoes a slow spontaneous decomposition into glutathione and lactic acid, but the decomposition is accelerated by glyoxalase. A different product is formed from methyl glyoxal and glutathione without the intervention of glyoxalase, and this decomposes spontaneously into the original substances.

We may for the sake of uniformity consider that glutathione is the prosthetic group of glyoxalase in the same sense that co-enzyme I is the prosthetic group of pyridinoproteins. Glutathione is acting catalytically in collaboration with glyoxalase presumably by undergoing a cycle of combination with methyl glyoxal and then regeneration to the original form. What is not clear is the mechanism whereby the internal oxido-reduction takes place.

Fig. 21 shows the dependence of the rate of conversion of methyl glyoxal to lactic acid on the concentration of —SH glutathione. The disulphide form of glutathione, cysteine and the pyridine nucleotides are inactive as coenzyme for glyoxalase.

Ascorbic oxidase

Plants and fruits which lack the polyphenol oxidase are usually found to be rich in peroxidase, an enzyme which catalyses the oxidation of phenols and amines by H_2O_2. Szent-Györgyi was struck with the fact that although purified peroxidase catalyses the immediate oxidation of phenols by H_2O_2, the freshly expressed sap of plants rich in peroxidase shows a decided lag before the coloured products of oxidation can be seen (35). Analysis of this discrepancy led to the discovery that some substance in the fresh sap was responsible for the lag period. This substance was isolated and shown to have the composition $C_6H_8O_6$. The original name given was hexuronic acid; the accepted chemical name is now ascorbic acid.

This substance is characterized by its ability to undergo reversible oxidation and reduction. Ascorbic acid reacts rapidly with the orthoquinones formed by the catalytic oxidation of phenols by oxygen. The lag period in freshly expressed sap is simply the time required for all the ascorbic acid to be oxidized by the orthoquinones. With exhaustion of ascorbic acid, the *ortho*quinones oxidize further to coloured decomposition and polymerization products (cf. Tauber (36)).

The Haworth and Reichstein schools have proved that ascorbic acid has the following constitution:

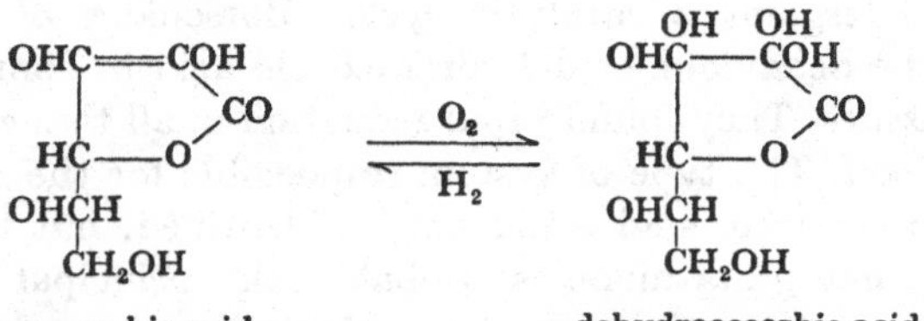

OHC═══COH   OHC——COH (with OH OH above)
  | \CO | \CO
HC——O/   HC——O/
  | |
OHCH   OHCH
CH₂OH   CH₂OH

ascorbic acid   dehydroascorbic acid

(O_2 / H_2 reversible arrows between structures)

In metal-free solution ascorbic acid is resistant to oxidation. Traces of metals, copper particularly, catalyse the oxidation by molecular oxygen (Barron *et al.* (37)). Tissue extracts inhibit the metalcatalysed oxidation almost completely and it is well established that tissue proteins and glutathione are the principal stabilizing agents. The oxidation product—dehydroascorbic acid—is not stable above pH 5. In the neutral or slightly alkaline range of pH its destruction is a matter of a few minutes (Ball (38)). Whether this instability applies *in vivo* is unknown, but it is unlikely that such would be the case.

There are three mechanisms known for the oxidation of ascorbic acid in plants: (1) ascorbic oxidase, discovered and characterized by Szent-Györgyi, (2) polyphenol oxidase and (3) peroxidase. Szent-Györgyi (39), Hopkins (40) and others have described ascorbic oxidase, an enzyme fairly widely distributed in plants (but not in animal tissues) which catalyses the oxidation of ascorbic acid by molecular oxygen. Plant peroxidase and H_2O_2 have also been shown to oxidize ascorbic acid in presence of traces of flavones as catalyst (41). Ascorbic acid reduces the the flavone; reduced flavone in turn is oxidized by the peroxidase-H_2O_2 system. These flavones are normal constituents of plants and their catalytic properties may therefore have physiological significance.

Szent-Györgyi has postulated that ascorbic acid plays the same role in plants as cytochrome in animal tissues. However, although mechanisms are known for the oxidation of ascorbic acid, little is known of the systems which reduce dehydroascorbic acid. It is premature therefore to arrive at any conclusions concerning the role of ascorbic acid in plants.

In animal tissues the vitamin invariably occurs in the reduced form. In view of this fact it is difficult to believe that ascorbic acid is undergoing a catalytic cycle. Borsook *et al.* (42) have studied the behaviour of dehydroascorbic acid in contact with animal tissues. They found rapid reduction in all tissues studied except blood. The type of system responsible for the reduction of dehydroascorbic acid could not be identified, but there was evidence that glutathione is probably the principal reducing agent. Stotz *et al.* (43) and Keilin & Hartree (44) have found that ascorbic acid is oxidized by molecular oxygen in presence of the cytochrome oxidase system. Tests of the catalytic action of ascorbic acid on dehydrogenase systems *in vitro* have been negative. There is no indication that ascorbic acid can act as a catalytic link between metabolites and oxygen.

The ascorbic oxidase catalyses the oxidation of ascorbic acid to dehydroascorbic acid by molecular oxygen. The expressed juice of cabbage and cauliflower leaves is rich in this enzyme. All varieties of *Brassica* contain the enzyme.

The enzyme works best at pH 6·0. The rate of oxidation of ascorbic acid is apparently independent of the concentration. That is to say, with a limited amount of ascorbic acid present, the rate of oxidation is constant until all the substrate has been oxidized. The Km is therefore extremely small. $M/1000$ cyanide inhibits the oxidation completely. Oxygen cannot be replaced by methylene blue or other acceptors.

Ascorbic oxidase does not catalyse the oxidation of reduced glutathione (GSH). Hopkins & Morgan[40] have shown that if both ascorbic acid and GSH are present together ascorbic acid is wholly protected against oxidation, whilst GSH is oxidized at exactly the same rate as ascorbic acid in absence of GSH. The oxidation of ascorbic acid promptly begins when virtually all the GSH present is oxidized (cf. Fig. 22).

Crook & Hopkins[45] have established that the induced oxidation of GSH involves an enzyme other than the ascorbic oxidase. This enzyme catalyses the reduction of dehydroascorbic acid by GSH. The mechanism of the induced oxidation of GSH may be represented as follows:

$$\text{ascorbic acid} + \tfrac{1}{2}O_2 \xrightarrow[\text{oxidase}]{\text{ascorbic}} \text{dehydroascorbic acid}$$

$$\text{dehydroascorbic acid} + \text{GSH} \rightarrow \text{ascorbic acid} + \text{GSSG}$$

The oxidation of GSH therefore depends only indirectly on ascorbic oxidase.

The ascorbic oxidase catalyses the oxidation of araboascorbic acid, glucoascorbic acid, galactoascorbic acid as well as of ascorbic acid[46]. Some specificity is shown in a preference for the laevo-rotatory members of the series. But the specificity is not complete.

Recently it has been reported (47) that the ascorbic oxidase preparation has no more catalytic activity than a solution of unspecific protein containing an equal amount of copper.

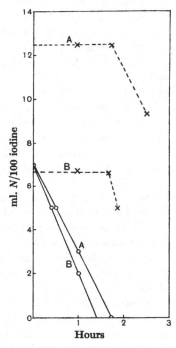

Fig. 22. The protective action of ascorbic acid on the oxidation of gluta-thione by the ascorbic oxidase system. The dotted lines represent the rate of oxidation of ascorbic acid in presence of glutathione whereas the continuous lines represent the rate of oxidation of glutathione in pre-sence of ascorbic acid. In case A the concentration of ascorbic acid was *ca.* twice that in B; that of GSH the same in each case. The iodine titre is a measure of the concentration of both ascorbic acid and GSH. pH 7·4, 18°. (Hopkins & Morgan [1936]. *Biochem. J.* **30**, 1446.)

A mixture of egg albumen and traces of copper was shown to simulate all the observed properties of the ascorbic oxidase. This interpretation appears somewhat extreme in view of the thermo-lability, partial optical specificity and fragility of the ascorbic oxidase—properties which strongly suggest a specific protein catalyst. There are some suggestions and hints in the literature that ascorbic oxidase may be a copper protein (cf. Ramasarma *et al.* (58) and also the recent isolation of the enzyme from squash as described on p. 118).

Aldehyde oxidase (potato)

The expressed juice of potato contains an enzyme which catalyses the oxidation of aldehydes by nitrate, methylene blue and other hydrogen acceptors but not by molecular oxygen (cf. Bach (59), Michlin (50), Bernheim (51)). A systematic investigation of the potato enzyme has been carried out by Bhagvat (52). The following is her method of preparation. The expressed juice of potatoes is treated at 0° with an equal volume of acetone. The precipitate is washed with cold acetone and then extracted with 10 % NaCl. The solution is made 40 % saturated with respect to $(NH_4)_2SO_4$ and the precipitate is redissolved in water. The enzyme precipitates when the solution is dialysed free of salts and after centrifuging off is redissolved in 10 % NaCl. The enzyme solution is further purified by treatment with charcoal. Such a preparation is *ca.* 112 times more active than the crude press juice.

The enzyme catalyses the oxidation of aldehydes either aliphatic or aromatic. No coenzyme is necessary for the reaction. As oxidant, nitrate or oxidation-reduction indicators such as methylene blue are equally effective (Bernheim (51)), although the kinetics are to some extent dependent on the chemical nature of the oxidant (Bhagvat (52)).

The enzyme is irreversibly destroyed when shaken in air in presence of some aldehyde. In absence of substrate no destruction occurs. The destruction has been shown to be due to peroxide formation. $M/300$ HCN inhibits the enzyme almost completely. Incubation with dilute cyanide, $M/1000$, reversibly inactivates the enzyme.

Carotene oxidase

Sumner & Dounce (56) have called attention to an enzyme present in different types of beans which catalyses the oxidation of carotene and other unsaturated fats by molecular oxygen. The existence of this enzyme has long been known in the milling industry, though it was never described in the scientific literature until the note of Sumner & Dounce appeared. The enzyme attacks carotene (mixture of α and β), xanthophyll, olive oil, cotton-seed oil, ethyl oleate and linseed oil, but has no action on saturated fats. The oxidation probably involves the addition of oxygen at a double bond with formation of the corresponding peroxide. This process of peroxide formation involves bleaching of the colour of the unsaturated hydrocarbon.

Hydrogenase

Stephenson & Stickland[57] described an enzyme in various bacteria which catalyses the reduction of oxygen, nitrate, sulphate, fumarate and other hydrogen acceptors by molecular hydrogen. A cell-free preparation of the enzyme has been obtained from suspensions of *Bact. coli* crushed in the Booth-Green mill, though as yet no study of its properties has been made.

Hydrogenase, like platinum black, catalyses the following reversible reaction:

$$H_2 \rightleftharpoons 2H \rightleftharpoons 2H^+ + 2\epsilon$$

Green & Stickland[61] have shown that the same equilibrium point between molecular and ionic hydrogen is reached in the presence of the bacterial enzyme as in the presence of platinum black.

REFERENCES

(1) Weil-Malherbe [1937]. *Biochem. J.* 31, 2080.
(2) Hare [1928]. *Biochem. J.* 22, 968.
(3) Pugh & Quastel [1937]. *Biochem. J.* 31, 286.
(4) Blaschko, Richter & Schlossmann [1937]. *Biochem. J.* 31, 2187.
(5) Pugh & Quastel [1937]. *Biochem. J.* 31, 2306.
(6) Richter [1937]. *Biochem. J.* 31, 2022.
(7) Philpot [1937]. *Biochem. J.* 31, 856.
(8) Batelli & Stern [1909]. *Biochem. Z.* 19, 219.
(9) Keilin & Hartree [1936]. *Proc. Roy. Soc.* B, 119, 114.
(10) Davidson [1938]. *Nature, Lond.*, 141, 790; *Biochem. J.* 32, 1386.
(11) Holmberg [1939]. *Nature, Lond.*, 143, 604.
(12) Best & McHenry [1930]. *J. Physiol.* 70, 349.
(13) Edlbacher & Zeller [1937]. *Helv. chim. Acta*, 20, 717.
(14) Zeller [1938]. *Helv. chim. Acta*, 21, 880.
(15) Bernheim, Bernheim & Gillaspie [1936]. *J. biol. Chem.* 114, 657.
(16) Müller [1928]. *Biochem. Z.* 199, 136; [1929]. *Ibid.* 205, 111; 213, 211; [1931]. *Ibid.* 232, 423.
(17) Franke & Lorenz [1937]. *Liebigs Ann.* 532, 1.
(18) Green, Stickland & Tarr [1934]. *Biochem. J.* 28, 1812.

(19) Tarr [1938]. *Nature, Lond.*, 142, 1078.
(20) Banga & Szent-Györgyi [1938]. *Hoppe-Seyl. Z.* 255, 57.
(21) Banga & Phillippot [1938]. *Hoppe-Seyl. Z.* 258, 147.
(22) Theorell & Swedin [1939]. *Naturwissenschaften*, 27, 95.
(23) Braunstein & Kritzmann [1937]. *Enzymologia*, 2, 129; 2, 138; [1938]. *Biochimia*, 3, 602.
(24) Kritzmann [1938]. *Biochimia*, 3, 614.
(25) Kritzmann [1939]. *Nature, Lond.*, 143, 603.
(26) Lohmann [1932]. *Biochem. Z.* 254, 452.
(27) Hopkins [1929]. *J. biol. Chem.* 84, 269.
(28) Kendall, MacKenzie & Mason [1929]. *J. biol. Chem.* 84, 657.
(29) Meldrum & Dixon [1930]. *Biochem. J.* 24, 472.
(30) Hopkins & Elliott [1931]. *Proc. Roy. Soc.* B, 109, 58.
(31) Quastel & Wheatley [1932]. *Biochem. J.* 26, 2169.
(32) Giršavičius & Heyfetz [1935]. *Nature, Lond.*, 136, 645.
(33) Yamazoye [1936]. *J. Biochem., Tokyo*, 23, 319.
     Nagaya, Yamazoye & Nakamura [1936]. *J. Biochem., Tokyo*, 23, 41.
(34) Kohn [1937]. *Biochem. J.* 31, 1693.
(35) Szent-Györgyi [1928]. *Biochem. J.* 22, 1387.
(36) Tauber [1936]. *Enzymologia*, 1, 209.
(37) Barron, de Meio & Klemperer [1935]. *J. biol. Chem.* 112, 625.
(38) Ball [1937]. *J. biol. Chem.* 118, 219.
(39) Szent-Györgyi [1931]. *J. biol. Chem.* 90, 385.
(40) Hopkins & Morgan [1936]. *Biochem. J.* 30, 1446.
(41) St Huszák [1937]. *Hoppe-Seyl. Z.* 247, 239.
(42) Borsook, Davenport, Jeffreys & Warner [1937]. *J. biol. Chem.* 117, 237.
(43) Stotz, Harrer, Schultze & King [1938]. *J. biol. Chem.* 122, 407.
(44) Keilin & Hartree [1938]. *Proc. Roy. Soc.* B, 125, 171.
(45) Crook & Hopkins [1938]. *Biochem. J.* 32, 1856.
(46) Johnson & Zilva [1937]. *Biochem. J.* 31, 1366.
(47) Stotz, Harrer & King [1937]. *J. biol. Chem.* 119, 511.
(48) Holmberg [1939]. *Nature, Lond.*, 143, 604; *Biochem. J.* 33, 1901.
(49) Bhagvat, Blaschko & Richter [1939]. *Biochem. J.* 33, 1338.
(50) Michlin [1927]. *Biochem. Z.* 185, 216; [1928]. *Ibid.* 202, 329.
(51) Bernheim [1928]. *Biochem. J.* 22, 125.
(52) Bhagvat [1939]. Ph.D. Thesis, Cambridge University.
(53) Blaschko [1939]. *Nature, Lond.*, 145, 27.
(54) Mann & Quastel [1939]. *Nature, Lond.*, 144, 943.
(55) Swedin & Theorell [1940]. *Nature, Lond.*, 145, 71.
(56) Sumner & Dounce [1939]. *Enzymologia*, 7, 130.
(57) Stephenson [1928]. *Biochem. J.* 22, 605.
(58) Ramasarma, Datta & Doctor [1940]. *Enzymologia*, 8, 108.
(59) Bach [1913]. *Biochem. Z.* 52, 412.
(60) Zeller, Stern & Wenk [1940]. *Helv. chim. Acta*, 23, 3.
(61) Green & Stickland [1934]. *Biochem. J.* 28, 898.

CHAPTER IX

Oxidations in organized systems

Biological oxidations can be studied either by way of the analysis of organized cells or by way of the analysis of cell extracts and enzyme systems. As would be expected, the type of analysis possible becomes more limited as the organization becomes more complex. Thus beyond measurements of respiratory quotients and feeding experiments there is no direct approach to the study of tissue oxidations using the whole animal as experimental material. The method of organ perfusion gives somewhat more scope for analysis, in that substances can be supplied to the organ and their fate can be ascertained by suitable methods. Much of our present knowledge concerning the oxidation of fats in liver has been derived from perfusion experiments. The tissue slice is the simplest form of cellular organization which can be handled experimentally, and it is not surprising that the analysis of oxidative processes by the tissue-slice method has in a few years proved as fruitful as three decades of organ perfusion.

There are many transition stages from the organized cell to the enzyme extract. Finely minced brain, for example, may contain no semblance of gross cellular organization, although much of the chemical organization has suffered relatively little change. Similarly, washed muscle mince forms a type of intermediary link between the organized cell and the enzyme extract. It is impossible at the moment to state precisely what is meant by chemical organization. The existence of such organization must be postulated to explain why certain reactions which occur in minced tissues and cell-free systems cannot be reproduced in extracts.

There are some who consider that the study of isolated oxidation enzymes may be amusing as chemical exercises but that it has little relevance to the problem of physiological oxidations. On the other hand, there is a tendency for those who are engaged in reconstructions of cellular oxidations, to forget that the countless chemical permutations and combinations of which the isolated components of oxidation systems are capable, cannot be assigned physiological significance until some counterpart of these events is shown to take place in the intact cell and organism. It is becoming increasingly clear that the study of enzyme systems cannot be divorced from the study of the intact cell and vice versa. Study of the one without reference to the other becomes sterile.

As yet it is not always possible to reproduce in enzyme systems reactions observed in the intact cell. For example, liver slices attack certain lower fatty acids energetically. This power of oxidation is destroyed immediately the slice is minced. This may mean either that the oxidation of lower fatty acids depends on the integrity of the cell structure or that some component of the catalytic system is destroyed during the process of mincing the tissue. Whatever the correct explanation the fact remains that at the moment the study of the oxidation of lower fatty acids is not amenable to analysis outside the organized cell. Similarly, the oxidation of certain natural amino acids, the oxidation of hexose and the linkage of certain oxidations with the synthesis of substances such as glycogen, urea, glutamine, etc. are processes which have not been reconstructed in tissue extracts. The difficulties in the way of such reconstructions very likely will diminish in proportion to the increase in the technical skill of those engaged in the isolation of enzymes. Special methods will have to be developed for the handling of the fragile enzymes concerned in the oxidation of fatty acids, amino acids, etc.

The rest of this chapter will be devoted to a consideration of some experiments with tissue slices and mince which are relevant to the problem of the mechanisms of biological oxidations.

Succinic cycle

Szent-Györgyi[1,2,3] and his coworkers have advanced an attractive theory of the role of C_4 dicarboxylic acids in the respiration of animal tissues. The experimental evidence for this theory was obtained principally from a study of the respiration of minced breast muscle of pigeon. The oxygen uptake of the mince falls off gradually with time. In presence of either succinate, fumarate, malate or oxaloacetate (all of which are inter-convertible) this decline in the rate of respiration is checked. Stare & Baumann (4) have shown that a few γ of any of the above four dicarboxylic acids were sufficient to produce the stabilizing effect. The difference in oxygen uptake between the experimental with dicarboxylic acid and control without addition was much greater than could be accounted for by the complete oxidation of added dicarboxylic acid. This fact would indicate that the dicarboxylic acids were acting as catalysts and not as metabolites. The succinic enzyme which catalyses the oxidation of succinate to fumarate is inhibited by malonate (Quastel(36)). If the succinic enzyme is concerned in the catalytic effect of the C_4 dicarboxylic acids it follows that malonate should reverse the stabilizing effect of C_4 dicarboxylic acids. Such is indeed the case.

Szent-Györgyi et al. (5) have formulated the following catalytic cycle for the C_4 dicarboxylic acids:

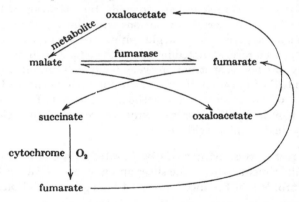

Oxaloacetate is reduced to malate at the expense of the oxidation of some metabolite such as triosephosphate. The rest of the cycle indicates how oxaloacetate is regenerated from malate. In presence of fumarase, malate is partially converted into fumarate.

$$COOHCHOHCH_2COOH \rightleftharpoons COOHCH : CHCOOH + H_2O$$

In presence of appropriate enzymes fumarate oxidizes malate to oxaloacetate and is thereby reduced to succinate. The cytochrome oxidase-oxygen system and the succinic dehydrogenase system collaborate in bringing about the aerobic oxidation of succinate to fumarate. The cycle then repeats. According to this scheme the oxaloacetic-malic-fumaric-succinic cycle acts in a catalytic capacity as intermediator between the metabolites on the one hand and the cytochrome-oxygen system on the other. It is thus possible for certain metabolities to become oxidized ultimately by the cytochrome system although unable to react directly with the cytochromes.

There are various points in connection with the Szent-Györgyi scheme which await clarification. Which metabolites are oxidized by oxaloacetate? There is some evidence that triosephosphate is the principal substance oxidized under Szent-Györgyi's experimental conditions (Banga[6], Greville[7]). The oxidation of triosephosphate by oxaloacetate would take place in a coenzyme I-linked reaction catalysed by the triosephosphoric and malic enzymes respectively. The oxido-reduction between fumarate and malate has been reconstructed in enzyme extracts by Green & Dewan[44] and shown to be a slow reaction. If this oxido-reduction as reconstructed is one of the stages in the catalytic cycle, the rapidity of the overall reaction is incompatible with the sluggishness of one of the component processes. The Szent-Györgyi scheme rests on the assumption that the succinic system is the only system in animal tissues which can react directly with the cytochromes. The available evidence is not in agreement with this assumption. Apart from the succinic and α-glycerophosphoric enzyme systems all the pyridinoproteins can react with the cytochromes indirectly via the Straub flavoprotein. But whatever the uncertainties as to the mechanism of the Szent-Györgyi cycle the catalytic effect of the C_4 dicarboxylic acids is an experimental fact which cannot be ignored and which ultimately must be allowed for in any comprehensive scheme of cellular respiration.

In presence of *Bact. coli* fumarate can oxidize glucose, malate, lactate, acetate, glycerol, glyceraldehyde, pyruvate, butyrate, $l(+)$-glutamate and molecular hydrogen (Quastel *et al.*[8,9], Krebs[10]). The rates of these anaerobic oxidations by fumarate were compared by Krebs with the rates of oxidation of the same substrates by molecular oxygen and were found to be of the same order of magnitude. The data were consistent with an earlier Szent-Györgyi view that the oxidation of certain metabolites by oxygen proceeds through the intermediation of the succinic-fumaric system, e.g.

metabolite + fumarate → oxidized metabolite + succinate
succinate + O_2 → fumarate

It must be remembered, however, that even if a particular oxidation can be shown to take place at the same rate in presence of fumarate as in presence of oxygen the conclusion does not necessarily follow that fumarate is the normal link between the substrate in question and oxygen.

Citric cycle

Krebs and his colleagues(11, 43) have obtained experimental evidence for a citric-acid cycle in the metabolism of certain animal tissues. Citrate catalytically promotes oxidations in muscle tissue, liver and kidney, especially in presence of carbohydrates. The effect is not observed in brain. The oxidation of citrate in muscle according to Krebs passes through the following stages:

| citric | cisaconitic | α-keto-glutaric | succinic | fumaric | malic | oxalo-acetic |

Pyruvic acid in presence of oxygen can condense with oxaloacetate to form citrate. In turn citrate is broken down to oxaloacetate. The net effect of the citric acid cycle is the complete oxidation of pyruvic acid. The individual reactions involved in the conversion of citrate to oxaloacetate have been studied by various investigators but not all in the same tissue. Martius & Knoop (12) and

Adler *et al.* (13) have shown that tissue extracts can catalyse the oxidative decarboxylation of citrate to α-ketoglutarate. Weil-Malherbe (14) has shown that the oxidative decarboxylation of α-ketoglutarate to succinate can take place in brain either aerobically or anaerobically. Under aerobic conditions for each molecule of α-ketoglutarate which is oxidized to succinate and CO_2, another is reduced to α-hydroxy-glutarate. Finally the succinic dehydrogenase, fumarase and the malic enzyme succeed one another in catalysing the oxidation of succinate to oxaloacetate. The only reaction still obscure is the condensation of pyruvate with oxaloacetate.

Strictly speaking it is isocitrate which is oxidized and not citrate (13). In presence of aconitase (12, 15, 16) citrate is partially converted first to *cis*aconitic acid and then to *iso*citric acid. According to Adler *et al.* (13) α-keto-β-carboxyglutaric acid is the unstable intermediary stage in the oxidation of *iso*citrate to α-ketoglutarate.

The Krebs citric cycle may be considered as a variant of the Szent-Györgyi C_4 theory in the sense that it embodies the conception of a metabolic cycle playing a catalytic role. But it should be pointed out that the Krebs cycle explains all the facts connected with the catalytic action of the C_4 dicarboxylic acids and raises none of the difficulties inherent in the Szent-Györgyi scheme.

Breusch (39) has quantitatively determined the production and disappearance of citrate in minced animal tissues. Kidney, and to a small extent liver, brain and lung condense oxaloacetate with pyruvate to give citrate. Muscle,* however, cannot either form citrate by the above condensation or utilize added citrate with any measurable velocity. In Breusch's view this fact rules out the application of the citric-acid cycle to muscle. As for kidney—the main site of citric-acid formation—Breusch considers citric-acid formation to be chiefly an elimination process which takes place only when the concentration of oxaloacetic

---

* Simola & Hollman (40) have found that pyruvate condenses with malate (but not with oxaloacetate) to form citrate in heart muscle. In other tissues oxaloacetate is more effective than malate. This anomaly may reconcile the findings of Krebs and Breusch.

acid exceeds some threshold value. These negative findings have been strongly criticised by Krebs[43].

Krebs & Eggleston[17] have investigated in great detail some of the factors which influence the respiration of minced pigeon-breast muscle. The falling off in oxygen uptake may be retarded by (a) boiled muscle extract, (b) citrate or substances derived from it in the citric acid cycle and (c) insulin. In presence of (a), (b) and (c) the most pronounced stabilizing effect is observed. The authors deduce from these facts that insulin acts as a catalyst in the citric-acid cycle. It is noteworthy that this is the first significant *in vitro* effect of insulin recorded in the literature. Shorr & Barker[37] were unable to obtain this insulin effect with minced muscle of laboratory animals other than pigeon and hold the view that this effect may not be concerned with the physiological action of insulin.

Oxidation of fatty acids

Our knowledge of the mechanisms whereby fatty acids are oxidized in the animal body has been derived mainly from liver-perfusion studies. It is beyond the confines of this book to deal with the classical work of Knoop, Friedemann, Dakin, etc. In recent years the direct analysis of fatty-acid metabolism in tissue slices has been initiated and the basis has been laid for an attack on the enzymes concerned.

Jowett & Quastel[18], Edson[19], Leloir & Munoz[20] and Cohen & Stark[21] have found that all normal saturated fatty acids (2–10 carbon atoms) are oxidized aerobically in liver. Acids of 4, 6 and 8 carbon atoms produce acetoacetic acid most rapidly, whereas the 10 carbon acid is slightly less active, and acetic acid considerably less. Acids of 5, 7 and 9 carbon atoms but not propionic produce acetoacetic acid at about $\frac{1}{3}$rd the rate of their even neighbours. Although the odd-numbered fatty acids yield proportionately less acetoacteic acid on oxidation the increased oxygen uptake (compared with control without substrate) which accompanies their oxidation is much greater than that for the even-numbered acids. Quastel & Jowett consider this fact as evidence that the oxidation of odd-numbered fatty acids is more complete. The unspecific methods at present available for estimation of β-ketonic acids leave room for doubt that acetoacetic acid is the only β-ketonic acid formed by the oxidation of fatty acids. There is some evidence that fatty acids with 5 or more C atoms give rise to an acid which is not identical either with acetoacetic or β-hydroxybutyric.

Jowett & Quastel(22) compared the aerobic rates of oxidation of crotonic and butyric acids in tissue slices under various conditions. The rates were practically identical in all cases. They concluded that the same catalyst is probably concerned in the two cases. The following is their scheme for the oxidation of butyric and crotonic acids in liver slices:

$$\left.\begin{array}{l} CH_3CH_2CH_2COOH \\ CH_3CH=CHCOOH \end{array}\right\} CH_3COCH_2COOH \rightleftharpoons CH_3CHOHCH_2COOH$$

According to this view β-hydroxybutyric acid is not an intermediate in the oxidation of the two fatty acids.

Leloir & Munoz(20) have made the interesting discovery that butyrate is oxidized aerobically in presence of fresh mince of rat liver. Both acetoacetate and β-hydroxybutyrate are formed as oxidation products. The mince is inactivated in a matter of minutes in absence of O_2. Succinate, fumarate, malate and citrate added to the mince increase the rate of disappearance of butyrate but exert no action once the mince has been inactivated by anaerobiosis. Attempts to isolate the enzyme system which oxidizes butyrate have thus far proved unsuccessful.

Acetate disappears rapidly in presence of liver and other tissues. As yet there are few clues as to the mechanism of acetate metabolism. The fact that the O_2 uptake of liver slices is increased in presence of acetate suggests that it is oxidized. Acetoacetate and β-hydroxybutyrate account for only a small part of the acetate disappearing. Glycollate and oxalate have been excluded as intermediates in the oxidative breakdown of acetate.

Krebs & Johnson(23) have made out a good case for the following mechanism of the formation of acetoacetate from acetate and pyruvate:

$$\text{acetate} + \text{pyruvate} \xrightarrow{-H_2O} \text{acetopyruvate}$$

$$\beta\text{-hydroxybutyrate} \underset{O_2}{\overset{H_2}{\rightleftharpoons}} \text{acetoacetate} \quad \downarrow -CO_2$$

Acetopyruvate has not been isolated from animal tissues, but when supplied to slices gives rise to acetoacetate or β-hydroxybutyrate at a rate even greater than that in presence of acetate and pyruvate. This postulated mechanism of Krebs & Johnson for the anaerobic disappearance of acetate supplies no explanation for the oxidative breakdown.

The Bernheims[24] have analysed the stimulating effect of vanadium on the respiration of rat or guinea-pig liver suspensions. They found that the substrate concerned is a phospholipin. In presence of vanadium, oxidation of the fatty-acid part of the phospholipin takes place. The amounts of vanadium necessary to induce the oxidation of phospholipin were approaching catalytic orders of magnitude (20–40γ vanadium per ml.).

Ciaranfi[38] claims that the methyl esters of monocarboxylic fatty acids are more rapidly oxidized in presence of tissue slices than the corresponding fatty acids. In liver, brain cortex and spleen the increase in oxygen uptake due to the esters are respectively 8, 11 and 17-fold greater than that due to the acids. The greater the carbon chain up to C_7 the greater is the rate of oxidation of the methyl ester.

Lang and his colleagues[42] have observed that finely divided and washed rat muscle or liver catalyses the oxidation of higher fatty acids particularly palmitic and stearic acids by methylene blue. Unsaturated acids show little or no activity. Stearic acid yields oleic acid on oxidation. Muscle adenylic acid and some of its derivatives increase the activity of the enzyme.

Aminophorases

Braunstein & Kritzmann[25] have discovered in animal tissues and plants a general series of reactions whereby NH_3 is transferred from an amino acid to a ketonic acid. The amino acid is oxidatively deaminated and the ketonic acid is reductively aminated. There is apparently a whole family of enzymes concerned in these transaminations. The glutamic and aspartic aminophorases have already been considered in the previous chapter.

The rules governing the transfer of NH_3 in minced pigeon-breast muscle may be summarized by the following diagram:

$$dicarboxy_{NH_2} \rightarrow \begin{cases} monocarboxy_{CO} \\ or \\ dicarboxy_{CO} \end{cases}$$

$$monocarboxy_{NH_2} \rightarrow dicarboxy_{CO} \rightarrow monocarboxy_{CO}$$

NH_2 = amino acid, CO = ketonic acid, and arrows indicate direction of transfer of NH_3.

The members of the dicarboxylic amino acids (e.g. glutamic) can transfer NH_3 to either monocarboxylic α-ketonic acids (e.g. pyruvic) or dicarboxylic α-ketonic acids (e.g. oxaloacetic). The monocarboxylic amino acids, however, cannot transfer NH_3 directly to the monocarboxylic α-ketonic acid except through the intermediation of catalytic amounts of some dicarboxylic α-ketonic acid. The dicarboxylic amino acids are therefore referred to as primary donators of NH_3, whereas the monocarboxylic amino acids are considered as secondary donators.

Glutamic and aspartic acids are the most effective donators of NH_3. According to Bychkov [26] the "availability for enzymatic NH_3 transfer appears to be due to the specific electrostatic configuration (polarity) of α-amino acids bearing two acidic groups at the ends of the molecule rather than to special chemical properties depending upon the presence of two carboxyl radicals". Thus phosphoserine, homocysteic and cysteic acids (pseudo-dicarboxylic amino acids in which phosphoric acid and sulphonic acid respectively provide the second acidic group) can act as NH_3 donators in transamination.

The monocarboxylic amino acids which act best as secondary NH_3 donators are lysine, phenylalanine, serine and leucine. There is some doubt whether glycine is active. Histidine and arginine are either slightly active or inactive. Direct transference of NH_3 from any of the above secondary NH_3 donators to pyruvic acid is not possible. In presence of catalytic amounts of dicarboxylic amino acids or their ketonic acids the transference goes on. The dicarboxylic amino acids and the corresponding ketonic acids act catalytically by undergoing a cycle of oxidative deamination by pyruvic acid (acceptor) and then of reductive amination by the monocarboxylic amino acid (donator). Thus in the transamination between phenylalanine and pyruvic acid catalysed by glutamic acid, the reaction takes place in the following stages:

pyruvic + glutamic → alanine + α-ketoglutaric

α-ketoglutaric + phenylalanine → glutamic + phenylpyruvic

It is clear from the above that the catalyst can be effective whether supplied as the amino acid or as the corresponding ketonic acid. Phosphoserine, cysteic and homocysteic acids cannot replace either glutamic or aspartic acids as catalytic carriers in transaminations between monocarboxylic amino and ketonic acids. Under optimum conditions each molecule of catalyst is aminated and deaminated 5–12 times per minute (T.N. = 5–12).

Amines and peptides cannot function as NH_3 donators. Aldehydes, ketones and hydroxy ketones cannot function as NH_3 acceptors. Mesoxalic acid, hydroxy-ketosuccinic and α-ketoadipic acid are however effective as NH_3 acceptors. Derivatives or homologues of pyruvic acid, such as α-ketobutyric, dimethyl pyruvic, methylethyl pyruvic and phenyl pyruvic acids, can replace pyruvic acid as NH_3 acceptor though not as efficiently. The corresponding amino acids of the above acceptors do not occur in nature. Both the d- and l-stereochemical series of amino acids are transaminated though the enzymes involved are specific for one or the other stereochemical form of the amino acid in question.

Cohen (27) has carried out experiments on pigeon-breast muscle which give information as to the quantitative importance of transamination. Significant transamination was found to take place only with alanine, glutamic, aspartic and their corresponding ketonic acids. Cysteine, α-amino-butyric, valine, phenylalanine and tyrosine were attacked at one-third to one-tenth the rate of the above group. All other amino acids were found to be inactive.

Energetic coupling of oxidation processes

Oxidation processes provide energy for carrying on the normal functions of the cell. Energy of oxidation is converted into other forms, such as electrical, mechanical, thermal, osmotic, etc. The mechanisms involved in these transformations of energy are totally obscure. Some beginnings have been made towards an understanding of how different types of chemical processes can be linked energetically. We shall consider here a few experiments bearing on the general problem of energy transformation.

According to Krebs [28] the synthesis of urea in mammalian liver takes place by combination of ornithine with two molecules of NH_3 and one of CO_2 to form arginine, which in turn is fragmented into ornithine and urea. Anaerobically synthesis of urea does not take place. This observation seems somewhat surprising at first sight, in view of the fact that none of the reactions involved in the synthesis of urea depend upon the presence or absence of oxygen. Reference to the thermodynamic data [29] shows, however, that the synthesis of urea from NH_3 and CO_2 requires considerable energy. Clearly the dependence of the synthesis on aerobic condition argues that some aerobic process supplies energy for the synthesis. Significantly glucose, fructose, lactate and pyruvate all increase the rate of urea synthesis some 200 %. Amino acids—such as alanine and aspartic acids—also stimulate the rate of urea synthesis. The effects are of the same order of magnitude as those produced by lactate or pyruvate and probably involve similar mechanisms, as shown by the fact that amino acids exert no effect when the solutions already contain lactate or pyruvate. The most plausible interpretation of these facts is that the synthesis of urea is linked energetically with the oxidation of any one of a limited group of metabolites.

Glutamic acid in presence of brain, retina and kidney can react with NH_3 to form glutamine (Krebs [30]). The hydrolysis of glutamine to glutamic acid and NH_3 goes to completion in presence of a specific enzyme. Thus energy must be supplied to the system in order to reverse the hydrolysis. Krebs has found that the synthesis of glutamine is linked with some aerobic oxidation process in kidney cortex and brain. HCN inhibits the synthesis of glutamine to approximately the same degree as it inhibits respiration. Under anaerobic conditions no synthesis takes place. In retina, however, the presence or absence of oxygen has no influence on the synthesis of glutamine. According to Krebs the anaerobic glycolytic system, which is very active in retina, supplies the energy for the synthesis. Immediately the slice is minced all capacity to synthesize glutamine is lost. The same applies to the synthesis of urea.

Needham & Pillai (31) and Meyerhof *et al.* (32) have studied the synthesis of adenosinetriphosphate from adenylic acid and phosphate, which accompanies the oxidation of triosephosphate by coenzyme I in muscle extracts. Reagents which inhibit the oxidation of triosephosphate prevent the synthesis to a corresponding degree. For each molecule of triosephosphate oxidized one molecule of adenylic acid is phosphorylated. In presence of arsenate oxidation of triosephosphate is not accompanied by synthesis of adenosinetriphosphate. The connection between the oxidation and the phosphorylation was clarified by the discovery of Negelein & Brömel (35) and Warburg & Christian (41) that phosphate is directly concerned in the oxidation of triosephosphate. In presence of the appropriate enzyme 3-glyceraldehydephosphate combines with phosphate, and the resulting diphosphoglyceraldehyde is then oxidized to 1 : 3- diphosphoglycerate. The latter can transfer phosphate to adenylic acid in presence of an enzyme present in skeletal muscle extract. The arsenophospho compound corresponding to diphosphoglycerate apparently cannot phosphorylate adenylic acid. Hence the inhibiting effect of arsenate in the coupled phosphorylation.

Lipmann (33) has discovered that the oxidation of pyruvic acid in presence of *Bact. Delbruckii* can be coupled with the synthesis of adenosinetriphosphate from adenylic acid and inorganic phosphate. The coupling take splace anaerobically or aerobically. That is to say, the coupling is independent of the nature of the oxidant for pyruvic acid. It appears that pyruvic acid becomes phosphorylated and simultaneously oxidized to acetylphosphate and CO_2 (Lipmann (34)). Acetylphosphate then acts as phosphate donator for adenylic acid in presence of appropriate enzymes.

We must await more data to decide whether the phosphorylation of adenylic acid by acetylphosphate can proceed independently of the oxidation of pyruvic acid.

REFERENCES

(1) Gözsy & Szent-Györgyi [1934]. *Hoppe-Seyl. Z.* 224, 1.
(2) Annau, Banga, Huszák, Laki, Straub & Szent-Györgyi [1935]. *Hoppe-Seyl. Z.* 236, 1.
(3) Annau, Banga, Blazsó, Bruckner, Laki, Straub & Szent-Györgyi [1936]. *Hoppe-Seyl. Z.* 244, 8.
(4) Stare & Baumann [1936]. *Proc. Roy. Soc.* B, 121, 338.
(5) Laki, Straub & Szent-Györgyi [1937]. *Hoppe-Seyl. Z.* 247, 1.
(6) Banga [1937]. *Hoppe-Seyl. Z.* 249, 205.
(7) Greville [1937]. *Biochem. J.* 31, 2274.
(8) Quastel, Stephenson & Whetham [1925]. *Biochem. J.* 19, 304.
(9) Quastel & Wooldridge [1929]. *Biochem. J.* 23, 115.
(10) Krebs [1937]. *Biochem. J.* 31, 2095.
(11) Krebs & Johnson [1937]. *Enzymologia*, 4, 148.
(12) Martius & Knoop [1937]. *Hoppe-Seyl. Z.* 246, 1.
     Martius [1939]. *Hoppe-Seyl. Z.* 257, 29.
(13) Adler, Euler, Günther & Plass [1939]. *Biochem. J.* 33, 1028.
(14) Weil-Malherbe [1937]. *Biochem. J.* 31, 2202.
(15) Johnson [1939]. *Biochem. J.* 33, 1046.
(16) Martius [1938]. *Hoppe-Seyl. Z.* 257, 29.
(17) Krebs & Eggleston [1938]. *Biochem. J.* 32, 913.
(18) Jowett & Quastel [1935]. *Biochem. J.* 29, 2159.
(19) Edson [1935]. *Biochem. J.* 29, 2082.
(20) Leloir & Munoz [1939]. *Biochem. J.* 33, 734.
(21) Cohen & Stark [1938]. *J. biol. Chem.* 126, 97.
(22) Jowett & Quastel [1935]. *Biochem. J.* 29, 2143.
(23) Krebs & Johnson [1937]. *Biochem. J.* 31, 772.
(24) Bernheim & Bernheim [1939]. *J. biol. Chem.* 127, 353.
(25) Braunstein & Kritzmann [1937]. *Enzymologia*, 2, 129; [1938]. *Biochimia*, 3, 602; [1939]. *Ibid.* 4, 303.
(26) Bychkov [1939]. *Biochimia*, 4, 189.
(27) Cohen [1939]. *Biochem. J.* 33. In press.
(28) Krebs & Henseleit [1932]. *Hoppe-Seyl. Z.* 210, 33.
     Krebs [1934]. *Ergebn. Enzymforsch.* 3, 247.
(29) Borsook [1933]. *Proc. nat. Acad. Sci.* 19, 627; *Ibid.* 19, 720.
(30) Krebs [1935]. *Biochem. J.* 29, 1951.
(31) Needham & Pillai [1937]. *Biochem. J.* 31, 1837.
(32) Meyerhof, Schulz & Schuster [1937]. *Biochem. Z.* 293, 309.
(33) Lipmann [1939]. *Nature, Lond.,* 143, 281.
(34) Lipmann [1939]. *Nature, Lond.,* 144, 381.
(35) Negelein & Brömel [1939]. *Biochem. Z.* 301, 135.
(36) Quastel [1926]. *Biochem. J.* 20, 166.
(37) Shorr & Barker [1939]. *Biochem. J.* 33, 1798.
(38) Ciaranfi [1939]. *Nature, Lond.,* 144, 751.
(39) Breusch [1939]. *Biochem. J.* 33, 1757.

(40) Simola & Hollman [1939]. *Science*, 90, 594.
(41) Warburg & Christian [1939]. *Biochem. Z.* 303, 40.
(42) Lang & Mayer [1939]. *Hoppe-Seyl. Z.* 261, 249; 262, 120.
   Lang & Adickes [1939]. *Hoppe-Seyl. Z.* 262, 123.
   Lang [1939]. *Hoppe-Seyl. Z.* 261, 240.
(43) Krebs & Eggleston [1940]. *Biochem. J.* 34, 442; Krebs [1940]. *Ibid.*
   34, 460.
(44) Green & Dewan [1937]. *Biochem. J.* 31, 1069.

INDEX

Printed in the United States
By Bookmasters